Children's Health and Wellbeing in Urban Environments

How children experience, negotiate and connect with or resist their surroundings impacts on their health and wellbeing. In cities, various aspects of the physical and social environment can affect children's wellbeing. This edited collection brings together different accounts and experiences of children's health and wellbeing in urban environments from majority and minority world perspectives.

Privileging children's expertise, this timely volume explicitly explores the relationships between health, wellbeing and place. To demonstrate the importance of a place-based understanding of urban children's health and wellbeing, the authors unpack the meanings of the physical, social and symbolic environments that constrain or enable children's flourishing in urban environments. Drawing on the expertise of geographers, educationists, anthropologists, psychologists, planners and public health researchers, as well as nurses and social workers, this book, above all, sees children as the experts on their experiences of the issues that affect their wellbeing.

Children's Health and Wellbeing in Urban Environments will be fascinating reading for anyone with an interest in cultural geography, urban geography, environmental geography, children's health, youth studies or urban planning.

Christina R. Ergler is a Lecturer in Social Geography at the University of Otago, New Zealand.

Robin Kearns is a Professor of Geography in the School of Environment at the University of Auckland, New Zealand.

Karen Witten is a Professor of Public Health at Massey University, New Zealand.

Geographies of Health
Series Editors

Allison Williams, Associate Professor, School of Geography and Earth Sciences, McMaster University, Canada

Susan Elliott, Professor, Department of Geography and Environmental Management and School of Public Health and Health Systems, University of Waterloo, Canada

There is growing interest in the geographies of health and a continued interest in what has more traditionally been labeled medical geography. The traditional focus of 'medical geography' on areas such as disease ecology, health service provision and disease mapping (all of which continue to reflect a mainly quantitative approach to inquiry) has evolved to a focus on a broader, theoretically informed epistemology of health geographies in an expanded international reach. As a result, we now find this subdiscipline characterized by a strongly theoretically-informed research agenda, embracing a range of methods (quantitative, qualitative, and the integration of the two) of inquiry concerned with questions of risk, representation and meaning, inequality and power, culture and difference, among others. Health mapping and modeling has simultaneously been strengthened by the technical advances made in multilevel modeling, advanced spatial analytic methods and GIS, while further engaging in questions related to health inequalities, population health, and environmental degradation.

This series publishes superior quality research monographs and edited collections representing contemporary applications in the field; this encompasses original research as well as advances in methods, techniques, and theories. The *Geographies of Health* series will capture the interest of a broad body of scholars within the social sciences, the health sciences, and beyond.

Also in the series

Geographies of Plague Pandemics (forthcoming)
The Spatial-Temporal Behavior of Plague to the Modern Day
Mark R. Welford

Non-Representational Theory and Health (forthcoming)
The Health in Life in Space-Time Revealing
Gavin J. Andrews

Environments, Risks and Health
Social Perspectives
John Eyles and Jamie Baxter

Children's Health and Wellbeing in Urban Environments

Edited by Christina R. Ergler,
Robin Kearns and Karen Witten

Routledge
Taylor & Francis Group
LONDON AND NEW YORK

First published 2017 by Routledge

2 Park Square, Milton Park, Abingdon, Oxon OX14 4RN
605 Third Avenue, New York, NY 10017

Routledge is an imprint of the Taylor & Francis Group, an informa business

First issued in paperback 2021

Publisher's Note

The publisher has gone to great lengths to ensure the quality of this reprint but points out that some imperfections in the original copies may be apparent.

British Library Cataloguing-in-Publication Data
A catalogue record for this book is available from the British Library

Library of Congress Cataloging-in-Publication Data
Names: Ergler, Christina R., editor. | Kearns, Robin A., 1959- editor. | Witten, Karen, editor.
Title: Children's health and wellbeing in urban environments/edited by Christina R. Ergler, Robin Kearns and Karen Witten.
Description: Abingdon, Oxon; New York, NY : Routledge, 2017. | Series: Geographies of health series | Includes bibliographical references and index.
Identifiers: LCCN 2016058391 | ISBN 9781472446015 (hardback) | ISBN 9781315571560 (ebook)
Subjects: LCSH: City children–Health and hygiene. | City children–Social conditions. | Urban health. | Child welfare.
Classification: LCC HT206.C453 2017 | DDC 305.2309173/2–dc23
LC record available at https://lccn.loc.gov/2016058391

ISBN: 978-1-4724-4601-5 (hbk)
ISBN: 978-0-367-21899-7 (pbk)

Typeset in Times New Roman
by Sunrise Setting Ltd, Brixham, UK

Contents

Illustrations

Figures

Tables

Boxes

Contributors

Paivi K. Abernethy, PhD MRes MSc, is currently working as Postdoctoral Research Associate at Royal Roads University in Victoria, Canada, and as Adjunct Professor at the University of Waterloo, Ontario, Canada. She is a former health promotion specialist in chronic disease prevention, Indigenous health and children's environmental health, and has published about bridging health and sustainability for children's wellbeing. Her current research projects focus on co-building sustainability and reconciliation, and co-production of knowledge for better policy development.

Stuart C. Aitken is Professor of Geography and June Burnett Chair at San Diego State University (SDSU), USA. He also directs the Center for Interdisciplinary Studies of Young People and Space at SDSU. Stuart's research interests include critical social theory, qualitative methods, children, youth, families, and communities. He is a past co-editor of *The Professional Geographer and Children's Geographies,* and his books include *The Ethnopoetics of Space and Transformation: Young People's Engagement, Activism and Aesthetics* (2014). Stuart has worked for the UN on issues of children's rights, migration, and dislocation.

Bree Akesson is an Assistant Professor at Wilfrid Laurier University's Faculty of Social Work, Waterloo, Ontario, Canada. Her research broadly focuses on global child protection and specifically on children and families affected by war, disaster, migration and poverty. Her current research examines the experiences of Syrian refugee families living in Lebanon.

Hannah Badland conducts research focusing on examining associations between health behaviours and outcomes, the urban environment, and transport at the neighbourhood level in children and adults. In parallel she is developing and testing theoretical frameworks to identify ecological associations with social determinants of health, wellbeing and inequalities. She is a Principal Research Fellow at the Centre of Urban Research, RMIT University, Melbourne, and has previously worked in New Zealand and the UK.

Ann E. Bartos is a Lecturer in the School of Environment at the University of Auckland, New Zealand. Her research with youth centres around a longitudinal ethnographic project she began in 2008 in rural New Zealand. Focusing on embodiment, relationalities and the environment, her research explores the possibilities for young people's political agency, political ecologies and care.

Harriot Beazley is a Senior Lecturer in Human Geography at the University of the Sunshine Coast, Queensland, Australia, and is Commissioning Editor (Pacific) for the Routledge journal *Children's Geographies*. Harriot's research seeks to understand the experiences of children and young people through child-centred, rights-based participatory research in Southeast Asia (especially Indonesia and Cambodia).

Eugenia Brage is completing her PhD in Social Anthropology at the University of Buenos Aires, Argentina. Her doctoral research (funded by CONICET) analyses the migratory process of children with cancer from the northeast and northwest of Argentina to the capital of Buenos Aires in order to access diagnosis and medical treatment. She focuses on the gendered and social organisation of health care.

Penelope Carroll PhD is a Public Health Researcher at SHORE & Whariki Research Centre, Massey University, New Zealand. Her research interests include social policy, housing, neighbourhoods and health, community development and child-friendly cities. She has a background in journalism and communications and a strong interest in social justice.

Alison Carver is a Research Fellow at Australian Catholic University and an Honorary Fellow at Deakin University, Australia. Her PhD research examined how neighbourhood safety was related to youth physical activity, with particular emphasis on road safety, concern about strangers, perception of risk and subsequent restriction/modification of children's physical activity in their neighbourhood. Her postdoctoral research examined social and physical environmental factors that may influence children's active transport and independent mobility (i.e. their freedom to move around without adult accompaniment). Her current research will focus on the built environment and health among older adults.

Tara Coleman has a PhD in human geography from the University of Auckland, New Zealand, where she is a Professional Teaching Fellow in the Social Science for Public Health program. Tara's research interests encompass interdisciplinary approaches to health and wellbeing, feminist geographies, phenomenology, health policy and qualitative research methods.

Debra Flanders Cushing is a Lecturer and Postgraduate Coordinator in Landscape Architecture at Queensland University of Technology, Brisbane, Australia. Her research areas include intergenerational park design, healthy environments for healthy children and design pedagogy. Debra integrates research and teaching in order to design better communities for children and youth and to inspire future designers to practice evidence-based design.

Linor David has research interests in environmental health, equity and adult education. She completed her MEd at the University of Toronto, Canada, and works across Canada on developing community-based solutions to food insecurity and social exclusion.

Adrienne de Melo has a passion and expertise in developing and delivering education for sustainability programs for schools. Her research interests are in the impacts of the 'nature-deficit disorder' and ways to develop young people's connection to nature and environmental literacy. Adrienne is currently working at the Department of Conservation as an Outreach and Education Coordinator. Her role entails providing leadership and support to connect New Zealanders to the natural world and make conservation part of their everyday lives.

Kim N. Dirks is a university academic motivated by an interest in how urban living impacts on human health and how simple town planning choices and changes in human behaviour can lead to improved health and wellbeing. Her research focus is on air pollution and noise in relation to urban transport.

Christina R. Ergler is a Lecturer in Social Geography at the University of Otago, New Zealand. Her research interests are at the intersection of geography, sociology and public health, and they centre on interdisciplinary approaches to health and wellbeing, socio-spatial health inequalities, experiential dimension of health and wellbeing, and participatory research methods. Thematically she focuses on the links between play, transport and sustainable urban life.

Claire Freeman is a Professor in the Department of Geography, University of Otago, New Zealand, where she teaches in the Master of Planning Programme. Her interests are in environmental planning, including sustainable communities, planning for children and young people and planning with nature.

Lisa Gibbs is an Associate Professor and a Director of the Jack Brockhoff Child Health & Wellbeing Program at the University of Melbourne, Australia. She leads a range of community-based public health research studies examining social and environmental influences on child and family health and wellbeing. This includes studies addressing obesity prevention, social inclusion, oral health and disaster recovery. Her research focuses on engagement of marginalized or disadvantaged groups with an emphasis on community and policy outcomes achieved through university–community–government partnerships.

Alysa Handelsman is a Sociocultural Anthropologist trained at the University of Michigan and based in Guayaquil, Ecuador. Her research focuses on the everyday lives of children and families in Guayaquil's shantytowns. She is especially interested in children's understandings of and experiences with race, racism and socio-spatial segregation. She is committed to collaborative ethnographic research.

Robin Kearns is Professor of Geography in the School of Environment at the University of Auckland, New Zealand. His research interests span issues within social, cultural and health geography. He is an editor of *Health & Place*, and his most recent book is *The Afterlives of the Psychiatric Asylum* (2015).

Peter Kraftl is Professor of Human Geography at the University of Birmingham, UK. He has published five books and over 60 journal articles and book chapters on children's geographies, geographies of education and architecture.

Lisa Law is a Cultural Geographer with interests in the relation between people, place and knowledge/power – mostly in Southeast Asia, but also in tropical Australia. She is Associate Professor at James Cook University Cairns, Australia, where she researches tropical urban design. Lisa is the current Editor-in-Chief of the journal *Asia Pacific Viewpoint*.

Diana Lewis is a Mi'kmaq scholar from Sipekne'katik First Nation in Nova Scotia. She has a Master of Resource and Environmental Management degree and is completing her PhD in Sociology and Social Anthropology at Dalhousie University, Canada, where she is on faculty as the Coordinator of the Indigenous Studies Program. Her areas of expertise include environmental health, health impact assessment, community-based participatory research, Indigenous methodology, quantitative methods and resource management.

Shanon Lim is a Masters graduate from the University of Auckland, New Zealand. His research specializes in personal exposure to air pollution in urban settings, particularly on public transport. His research examines ways to reduce air pollution exposure for commuters by improving urban planning on transportation pathways.

Colin MacDougall is a Public Health Researcher concerned with equity, ecology and healthy public policy. He is Professor of Public Health at Flinders University, Australia, and an Executive Member of the Southgate Institute of Health, Society and Equity. A major research interest is exploring how children experience and act on their worlds, mainly with the Jack Brockhoff Child Health & Wellbeing Program at the University of Melbourne, where he holds an honorary position.

Jeffrey R. Masuda PhD is Associate Professor in the School of Kinesiology and Health Studies at Queen's University, Canada, and Canada Research Chair (Tier II) in Environmental Health Equity. His present work uses arts-based research methods to historicize the production of place-based health inequities and concomitant social justice activism in Vancouver, British Columbia.

K. Milam Brooks is a Doctoral Student in Educational Policy Studies at the University of Illinois at Chicago, USA. She received a Master's degree in Social Work from the same institution. Her research focuses on educational equality and the link between social supports and academic achievement among African American and Latino youth.

Bjorn Nansen is a Lecturer in Media and Communications at the University of Melbourne, Australia. He is a Fellow of the Melbourne Networked Society Institute and a member of the Microsoft Social NUI Research Centre and Research Unit in Public Cultures. His research interests include technology adoption, home media environments, young children's digital media use and post-digital interfaces. He currently holds an Australian Research Council-funded Discovery Early Career Researcher Award (DECRA) to study young children's use of mobile and interactive media.

Melody Oliver has a passion and expertise for understanding factors that encourage healthy lifestyles in children and their families. Her broad interests are in physical activity and health at a population and individual level, including the objective measurement of physical activity dimensions (habitual, transport-related, activity intensity, sedentary behaviors), and the integration of objective activity measurement and associated tools (e.g. pedometry, accelerometry, global positioning system (GPS) and GIS). She is particularly interested in physical activity and sedentary habits formed in childhood and how these can be influenced by children's social and built environments. She is an Associate Professor in the Faculty of Health Sciences at the University of Auckland, New Zealand.

Robin Quigg (*Raukawa*) is a Research Fellow in Cancer Society Social and Behavioural Research Unit at the University of Otago, New Zealand. Robin has broad and varied research interests and is particularly interested in Māori-centred, spatio-temporal epidemiology, using smart devices, spatial analysis, and geographic context to better understand health-related behaviours.

Pamela Anne Quiroz PhD (University of Chicago, 1993) is Director of the Center for Mexican American Studies and Professor of Sociology at the University of Houston, USA. Her areas of specialization are children and youth, identity, education, family and qualitative methods.

Jennifer A. Salmond is a Senior Lecturer in the School of Environment, University of Auckland, New Zealand, and specializes in urban meteorology and air pollution. Her research focuses on improving our ability to understand the temporal and spatial heterogeneity of pollutant concentrations in the urban atmosphere and their implications for human health.

Silvia Schäffer is a member of the GeoHealth Centre at the University of Bonn, Germany. She has worked for several projects concerning nature conservation and health. In 2017, she successfully completed her PhD entitled 'Nature experience and health: subjective health assessment, motor abilities and insights of forest kindergarten children'.

Michelle Thompson-Fawcett (*Ngāti Whātua*) is Associate Professor in Te Iho Whenua, Te Whare Wānanga o Ōtago, Aotearoa (New Zealand). Michelle's research focuses on power relations and practices of inclusion/exclusion in local planning. Her projects with Māori communities investigate processes of urban design, cultural landscape management, cultural impact assessment, and Indigenous resource management/planning.

Frank Vetere is Professor in the Department of Computing and Information Systems at the University of Melbourne, Australia. He is the Director of the Microsoft Research Centre for Social Natural User Interfaces (Social-NUI) and leads the Interaction Design Laboratory. Frank's expertise is in human–computer interactions (HCI) and social computing. His research aims to generate knowledge about the use and design of information and communication technologies (ICT) for human wellbeing and social benefit.

Karen Villanueva conducts research exploring associations between the neighbourhood-built environment and child health behaviours and outcomes. In particular, she is interested in locational and socio-environmental determinants of children's independent mobility, activity spaces, development and mental health. She is a Research Fellow at the Centre of Urban Research, RMIT University, Australia, and the Murdoch Children's Research Institute, Melbourne.

Cecilia Vindrola-Padros PhD is a Research Associate in the Department of Applied Health Research, University College London, UK. She received a PhD in Medical Anthropology from the University of South Florida, USA. She has carried out research on children's and young people's cancer care. She is particularly interested in understanding the experiences of families who travel long distances and relocate to access medical treatment.

Karen Witten is a Professor of Public Health at the SHORE & Whariki Research Centre, Massey University, New Zealand. Her research is focused on the way streets, neighbourhoods and cities are designed and used to promote or inhibit health and wellbeing.

Bronwyn E. Wood is a Senior Lecturer at the Faculty of Education, Victoria University of Wellington, New Zealand. Her research interests lie at the intersection of sociology, geography and education and they centre on issues relating to youth participation, citizenship and education. Her recent research focuses on the assessment of young people's social action in curriculum, restorative practices and civic life and affective educational spaces.

Helen Woolley has had two main research strands for over twenty years: strategic issues relating to Urban Open Spaces and how these are planned, managed and designed. Much of this work has fed into policy and practice for government, non-governmental organizations (NGOs) and charities in England. She also has a keen interest in children's outdoor environments from playgrounds, including the inclusion of disabled children, to the exclusion of skateboarders from civic urban open spaces. A focus on children's outdoor environments in post-disaster Japan has led to ongoing interests in other challenging contexts, including high density cities in China and refugee children in Turkey. Adults, institutions and society can support, facilitate and provide for—or control and constrain—children and young people in outdoor environments.

Preface

This book is the first 'play ground' in which health geographers and interdisciplinary colleagues discuss urban children's health and wellbeing between the covers of a dedicated volume. The idea for this collection arose at a Medical Geography Symposium in Durham, UK in 2011. In recent years an increasing number of health geographers have been inviting children onto centre-stage in their analyses; however, work has often been scattered across published outlets, limiting the voice of children beyond sub-disciplinary fields. In addition the recent and justifiable research focus on the implications of 'obesogenic environments' has potentially overshadowed the attention that many other pressing issues (e.g. poverty, pollution) for children growing up in urban environments deserve. Our aim therefore is to broaden the perspective on children's health and wellbeing.

Although our starting point for analysis—as health geographers—is always related to the links between health and place, in this collection we acknowledge the need for interdisciplinary perspectives to help create a better world with better health and wellbeing outcomes for children and future generations. In terms of vantage point, fifteen years ago Moon and Kearns termed health geography a 'magpie discipline' that seeks out the most helpful and alluring theoretical lenses. Similarly, we sought out an eclectic set of contributors from across the world to provide a partial snapshot of children's contemporary health and wellbeing issues. This work would not have been possible without all the children and adult participants who contributed their voices for this book and also our contributors who undertook to interpret these voices from a health geographic perspective as far as possible. All contributors responded with eagerness and rigor to our many queries and suggestions, for which we are grateful. As editors it has been interesting and fulfilling to work with scholars from such a diversity of disciplinary backgrounds and to help weave new connections into health geography.

We would like to thank Hayley Sparks for her thorough copy editing of the final manuscript and Chris Garden for his excellent cartographic work on the figures and tables. We also thank Katy Crossan and Priscilla Corbett for their guidance, and our respective institutions for support in undertaking this project.

Christina Ergler, Robin Kearns and
Karen Witten, 2016

Introduction

A children's place in health geography

Christina R. Ergler, Robin Kearns and Karen Witten

This book offers a critical perspective on primary school age children's health and wellbeing in urban places. We address the benefits of an interdisciplinary perspective through chapters that offer a series of 'windows' into children's worlds, variably emphasising theoretical and empirical vantage points, a range of scales and the opportunity to learn from diverse places.

Our intent has been to implicitly address a fundamental paradox that although children's wellbeing is seriously compromised in various ways across the world, children are able to express resilience and appear to find happiness – however fleeting – in unexpected ways and places. Experiences of resilience and happiness often occur despite the presence of a potent mix of influences that can encroach upon the freedoms of childhoods. Two examples from our 'backyard' are instructive. When two children were hit by a car near an Auckland school a member of the police was reported as saying it is '"unacceptable" for young children to be walking to school without adult supervision' (*NZ Herald*, 2015a). Earlier the same year in a new medium-density housing development in Auckland residents complained about the noise emanating from children's play equipment and a flying fox was removed by Auckland Council staff (*NZ Herald*, 2015b). These anecdotes are small events in global terms, but they speak to the challenges for children and their aspirations for outdoor play and independence.

Children seek agency in their everyday lives and yet they experience significant constraints shaped by societal structures, intergenerational relationships and urban environments designed to facilitate the lives of adults and not children. Despite these constraints, children do shape and transform their surrounding social, economic and political environments through mundane everyday actions. How they experience, negotiate and connect with (or resist) their surroundings impacts on their health and wellbeing.

In cities, multiple and interacting aspects of the physical and social environment can affect children's wellbeing. Around the globe, access to friends and places to socialise and play can foster children's social wellbeing and participation in community life whereas exposure to environmental pollution, violence and chronic poverty are more likely to undermine wellbeing (Hart *et al.*, 1999; Huo *et al.*, 2007; de Carvalho, 2013; Cardwell and Crighton, 2014). Children's social and mental wellbeing is also shaped by different family or parenting situations and practices (Pain, 2006; Taylor, 2009; Witten *et al.*, 2009; Talbot, 2013). In highly urbanised Western countries, trends such as increasing educational demands and engagement with new technologies, alongside declining independent exploration of local environments and loss of contact with 'nature', are changing children's attachment to places (Chawla, 1992; Freeman and Tranter, 2011; Ergler *et al.*, 2016). In brief, determinants of children's health and wellbeing in urban environments around the world are multifaceted and complex, encompassing diverse dimensions, structures and scales. Expressions of agency reflect this complexity as do children's experiences of place, which can range from the spatially constrained to the diffuse and virtually expansive.

Children actively participate in the construction of their family, school and community environments, but they are also constrained in these different settings by social, economic and political contexts. Childhoods are not universal and uniform. Geographical location and local and national histories cohere to shape the opportunities available to children in different places as well as their affective experiences of place (Philo, 2000; Ergler *et al.*, 2013a; Holloway, 2014). Although health and wellbeing have been implicitly addressed in recent publications in children's geographies and sociological studies of childhood (e.g. Freeman and Tranter, 2011; Hörschelmann and van Blerk, 2012; Bond, 2014), in this book we give these dimensions of children's everyday life priority.

This book also brings together accounts and experiences of children's health and wellbeing in urban environments from both majority and minority world perspectives. Where possible, we privilege children's expertise and explicitly explore the relationships between wellbeing and place. To demonstrate the importance of a place-based understanding of urban children's health and wellbeing our contributors unpack the meanings of the physical, social and symbolic environments that constrain or enable children's opportunities to flourish in urban environments. We are geographers. However, in the inclusive spirit of the subdiscipline of health geography, we highlight the importance of interdisciplinary perspectives in promoting an enhanced understanding of the diverse factors contributing to feeling well and being healthy in urban environments. We draw on the expertise of fellow geographers and also educationists, anthropologists, psychologists, planners and public health researchers along with nurses and social workers.

A place for children in health geography

We place this book at the intersection of health geography and children's geographies. The former subdiscipline grew out of the disease and health service-focused medical geography and evolved into a broader health-oriented identity following the mid-1990s (Kearns and Moon, 2002; Brown *et al.*, 2009). The latter field can be traced to calls for children's perspectives in an otherwise 'adultist' discipline (Winchester, 1991) twenty-five years ago. Both subdisciplines are now characterised by the fertile porosity of an interdisciplinary outlook. Adding ideas from environmental psychology and economics to pre-existing concerns with therapeutic places (Gesler, 1992), health geographers have increasingly embraced notions of wellbeing (Fleuret and Atkinson, 2007; Kearns and Andrews, 2010). Within other scholarly circles, Sarah James asked in the early 1990s whether there is a 'place for children in geography' (James and Prout, 1990: 1). In this collection we draw together these strands of research by addressing the place of children in health geography.

Our intentions in proposing this book were to privilege children's experiences of urban environments and acknowledge the various roles children have as social actors in their own right. We also sought to highlight the diversities of childhoods within and across different socio-cultural contexts (James and Prout, 1990; Holloway and Valentine, 2000; Holt, 2011). We share the contention of recent writers that children can voice their experiences and views on the world and contribute to change through child-centred methods (Christensen, 2000; van Blerk and Kesby, 2013; Ergler, 2015). Seeing children as valuable knowledge brokers on issues that impact on their health and wellbeing, however, has taken a long time to gain acceptance. As this collection affirms, this is a perspective that remains in its infancy in the subdiscipline of health geography.

Two helpful and mutually reinforcing trends raised the visibility of children in research and policy at the beginning of the 1990s. The United Nations Convention on the Rights of the Child (UNICEF, 1995) fostered the participation of children as full citizens in political, social and economic life. This agreement put children on the political agenda – or at least obliged policy makers to consult children on decisions that might affect their wellbeing (Percy-Smith and Thomas, 2010; Ergler, 2016). As a consequence, children have gained visibility within some political and policy arenas (Smith, 2016). Lawyers and judges consult children on their preferred living arrangements when parents separate (Taylor *et al.*, 2012). Doctors are obliged to inform and gain consent from children on medical procedures (Alderson, 2002; Donnelly and Kilkelly, 2011). Children and youth councils in school or local government settings have become a popular means of fostering children's participation in Western countries over recent decades (Matthews and Limb, 2003). An increasing number of local and national government agencies commission impact assessments of policies on children and their wellbeing (Mason and Hanna, 2009). Planners and architects have begun to design educational institutions, public places and leisure destinations with children (Freeman *et al.*, 1999; Clark, 2010; Derr and Tarantini, 2016).

At the same time and also reinforcing these developments, feminist and post-structural thinking gained currency in the research arena bringing marginalised voices to the fore. As a consequence of discussions around the politics of representation researchers began to work *with* children instead on their behalf (Holloway and Valentine, 2000; Barker and Weller, 2003; Cahill, 2007).

In keeping with these trends there has been a concurrent and growing recognition of children's status as independent social actors in many Western societies (James and Prout, 1990; Valentine, 2004) and increasingly in developing country contexts (Punch, 2002; Beazley, 2003; Percy-Smith and Thomas, 2010). This viewpoint positions children as more than invisible agents in institutions such as families or education facilities; instead it sees them as individuals with competence to create and transform their own social worlds within the structural constraints and enabling circumstances they encounter (James *et al.*, 1998; Christensen, 2000; Jones, 2001; Valentine *et al.*, 2001; Matthews, 2001; Kraftl *et al.*, 2007). Research is increasingly revealing the diversity of childhoods within and across places as well as differences in societal norms and rules that structure this diversity. As our contributors indicate, the way adultist society and power struggles over identities and resources structure children's lives remain central considerations in the quest to understand children's worlds and wellbeing (Matthews, 2001; Holt, 2004; Gallagher, 2008).

Placing children's health and wellbeing

In advocating a view of health as 'a state of complete mental, physical and social wellbeing' the World Health Organization (WHO) offered opportunities for addressing health and wellbeing in unprecedented ways. First, the WHO definition shifted the focus of attention to positive states of body and mind rather than sickness, ailment and illness. Health and wellbeing became intrinsically entangled, although geographical studies that address 'wellbeing' more prominently and explicitly only began to appear over the last decade (Fleuret and Atkinson, 2007; Kearns and Andrews, 2010; Atkinson *et al.*, 2012; Schwanen and Atkinson, 2015).

Second, the state of aspirational wellbeing advanced by the WHO definition is anchored in the social, economic, political, environmental and spatial contexts of everyday life. Viewed through an urban lens our bodies and minds can be seen as affected by not only the materialities of the cities we live in, but also the affective atmospheres generated by the less-than-fully-visible fault lines, such as discrimination and abuse. Not only does who we are matter, but where we live and what we do also influences our current and future health and wellbeing. By promoting a socio-ecological model of health, Dahlgren and Whitehead (1991) broadened understanding of the diverse determinants of health and wellbeing. The potential influences were extended beyond individual factors to acknowledge the effects of social and community networks, lifestyle and general living conditions (e.g. housing, service provisions, global environmental change). We note the prevailing place of parenting in mediating the tension of structure and agency dynamics in the lives of children (Witten *et al.*, 2015). To this extent, the (perhaps inevitable) involvement of parents in shaping the parameters of the lives of children means, as well as the potency and prevalence of expert knowledge, that adultist views on children's wellbeing prevail. While our own adult status and the distance – always temporal, sometimes spatial – from the spaces of our own childhoods is great, we have nonetheless sought where possible to include children's voices in the scholarship between the covers of this book.

Placing children's health and wellbeing in an urban environment

Early studies that engaged with children's wellbeing in urban environments reflect their position within the recent history of human geography more generally: the move from geography as a positivistic spatial science to geography as a fully *social* science emphasising subjectivity, experience and meaning of places. Two studies in the 1970s (Blaut and Stea, 1971; Bunge, 1973) paved the way for the explicit development of 'Children's Geographies' as a research field. Studies by Blaut and colleagues brought children into the research agenda of geographers through exploring children's experiences and mapping skills (Blaut and Stea, 1971; Blaut, 1987). This work researched children's spatial literacy in urban environments and initiated geographical research on children's developmental stages (Blaut, 1987; Blaut and Stea, 1971). Matthews (1985) followed in Blaut's footsteps and showed Piaget's misconception of children's development in relation to mapping abilities (Piaget and Inhelder, 1997). We view these early quantitative studies as a prerequisite for building a baseline for understanding children's own perception of their urban environment and environmental literacy.

In contrast, and inspired by the incipient social justice turn in geography, Bunge's (1977) 'geographical expeditions' were revolutionary both in their intent and in their methodology. Bunge literally and metaphorically undertook journeys into children's worlds in urban North America, applying qualitative methods to reveal young people's wellbeing in cities. His work highlighted inequalities in geographies of everyday life in deprived urban areas such as Detroit. Other researchers (Hart, 1979; Moore, 1986) followed his example in employing a humanistic approach when researching children's experiences and environmental literacy. These researchers drew upon the research methods of environmental psychology and ethnography (e.g. participant observation) to explore children's place experiences, their independent mobility, and their understanding of environmental processes and cognitive mapping abilities. Research examples that influenced the understanding of children's health and wellbeing in the widest sense include Ward's (1978) work on inner city slums in London in the 1970s, and Lynch's (1979) study of growing up in cities. These studies were highly descriptive, providing an in-depth perspective on children's environments. Nonetheless, children were unintentionally objectified in the research process. For instance, Bunge describes observing children's behaviour as 'similar to bird watching' (Bunge, 1973: 336). Although the aim of these studies was to gain insights into children's life worlds, the focus was less on children's voices per se and their understanding of wellbeing, and more on how adults interpret the impact of urban environments on children's health and wellbeing.

There is, however, a continuity between Bunge's 'expeditions' and our collection. Just as he sought to reveal the otherwise invisible contours of children's lives in places nearby but yet 'worlds apart', we also present some of the sites of childhood that lie 'under the radar' of the popular imagination and social policy (Hayes and Foster, 2002). Too often, it seems to us, marginal spaces of childhood are well beyond the spotlight of health promotion and urban liveability, and hence we include accounts of lesser-known places like forest kindergartens. We contend that much can be learnt from contexts that are marginal to our own (Western, privileged) experience but which, in their otherness to us, are nonetheless banal to many beyond our immediate horizon. Here we acknowledge, for instance, the suffering of those subjected to unrelenting violence in Ecuador and Palestine and whose sites and experiences of trauma are documented in this collection.

Children were 'seen but not heard' in early work by health and medical geographers. At the inaugural meeting of what became known as the International Medical Geography Symposium (IMGS) Gesler (1986) reviewed the use of spatial analysis in medical geography and discussed work by (non-geographer) Knox (1963) on space–time interactions in the occurrence of cleft palate in children. This work was distinguished by its abstract mathematical analysis and attention to a very particular ailment. An early medical-geographic contribution adopting an ecological perspective was published in one of the discipline's 'flagship' journals (Andrews, 1985).

It was arguably the social justice imperative of early work on inequalities that brought children more onto the agenda. Working at the interface with public health and medicine, for instance, Schneider and Greenberg at Rutgers published a number of papers concerned with child health (e.g. Schneider *et al.*, 1997). Such work was, however, largely written for and speaking to medical audiences. However, as medical geographers sought and found a more distinctive voice of their own, the very act of researching controversial children's health issues (e.g. effects of racism, violence) in public health was framed as a healthy activism (Greenberg *et al.*, 1990). In 1998 Rosenberg (1998) stated that 'Children in general are under-researched in medical and health geography'. Arguably it was the 'cultural turn' in health geography (Gesler, 1992: 216) that opened doors to more broadly framing the contexts of children's health and, in particular, the places and environments that shape wellbeing – ranging from therapeutic camps through hospitals to the walk to school (Kearns and Barnett, 2000; Kearns and Collins, 2000; Collins and Kearns, 2001; Collins and Kearns, 2005). Of note, however, these studies were of the *places* of the production of health and wellbeing for children and not of children's *experience* per se. It was only further into the next decade that methodological creativity and ethical soundness saw children and young people's voices appear in print to 'speak' to their experience (Mitchell *et al.*, 2007; Fenton *et al.*, 2013; Skinner and Masuda, 2013).

More recently, health geographers' engagement with children's health and wellbeing in urban environments has been sustained in contributing to debates around 'obesogenic environments' (Pearce and Witten, 2010). Studies have examined environmental determinants in relation to children's healthy levels of physical activity and independent mobility in schools, parks and neighbourhoods (Maddison *et al.*, 2009; Gilliland *et al.*, 2012; Doherty *et al.*, 2014; Smith *et al.*, 2015) and their food environments (Pearce *et al.*, 2009; Glen *et al.*, 2013; Vine and Elliott, 2013), teasing out aspects of neighbourhood context and composition. These debates helped influence the development of a child-related destination accessibility index (NDAI-C), which provides a comprehensive tool for assessing the accessibility and quality of different neighbourhood environments (Badland *et al.*, 2015). In brief, this line of research often utilises geographic information systems (GIS) combined with a global positioning system (GPS) to gather information about the contribution of environmental determinants on physical activity levels or food deserts (Loebach and Gilliland, 2016), an approach that leads to its own debates on accuracy and methodological improvements (Rainham *et al.*, 2010; Mavoa *et al.*, 2011; Oliver *et al.*, 2014). However, although better information about people's space-time behaviour and the relationship between environmental factors and health outcomes may be generated, this line of research is not able to reveal *how* children experience these environments and environmental influences. They run, as Andrews *et al.* (2012) argue, the risk of falling into the trap of environmental determinism.

As a response, 'qualitative geographical information systems' (qualitative GIS or SoftGIS) (Cope and Elwood, 2009) and also engaged qualitative studies using in-depth interviews and walking methods (e.g. Neuwelt and Kearns, 2006; Loebach and Gilliland, 2010; Carroll *et al.*, 2013) gained popularity for revealing nuanced understandings of children's experiences. Such approaches have also more recently addressed children's justification for so-called healthy or unhealthy practices in space and time (Ergler *et al.*, 2013b; McKendrick, 2014). These studies highlight the need for understanding healthy practices within the context of their production related to place, social positions and dominant societal discourses as a prerequisite for promoting and bringing healthy lifestyles into action. Evans (2006) and Evans and Colls (2009) advance a critique of discourse around healthy lifestyles by questioning the relevance of key measures used widely in public health (e.g. Body Mass Index) for its contribution to an implicit production of healthy citizenship in policy. Other researchers have shifted the focus from 'mainstream' children's issues towards marginalised groups, such as those facing barriers due to physical and mental health issues (e.g. Bingley and Milligan, 2007; Yantzi *et al.*, 2010). At the heart of all these studies is the quest to understand the social, cultural and environmental determinants for participation in active lifestyles in order to highlight the inequalities existing within and across different population groups and places. Researchers aim to inform and urge policymakers and planners at the local and national scale to tailor health promotional activities or urban design upgrades towards place-based needs of diverse social groups.

New stories of children's health and wellbeing

Our collection contributes to a new narration of the place of children in health geography. Children conventionally love stories, but as geographers, and social scientists more generally, we have been latecomers to acknowledging the power of stories to assist in understanding the determinants of health and wellbeing. Our concern is to tell a story of children's wellbeing that embraces subtle yet potent influences upon young lives (both tangible material structures and the 'affective atmospheres' described by Aitken in his concluding chapter).

To weave a new narrative of children's wellbeing, we privilege but are not limited to qualitative studies in this collection. One indication of new narratives in our book is the occurrence of metaphors, those applications of words or ideas to something that is imaginatively but not literally applicable to constructing understandings of changing places and states of wellbeing (Kearns, 1997). Examples of metaphors invoked by our contributors include media ecologies, companion devices, infectious happiness, alfresco children, 'park 'n' stride', the miasma of violence and therapeutic itineraries. In each case, authors have strategically chosen a term that, in its metaphorical content, is essentially malleable (which, as the dictionary tells us, is something 'able to be pressed into shape without breaking or cracking'). To take but one of the foregoing examples, Bree Akeson's use of the term 'miasma of violence' strategically name-checks the Victorian belief that pestilence travelled invisibly in the local atmosphere. Perhaps the somewhat audacious reclaiming of this outmoded term and its application to an 'air' of expectant if not actual violence effectively connects the plight of children to a lineage of suffering. It also offers urgency through deploying that word so often conveniently omitted from discussions of health status and wellbeing: violence.

Structure of the book

When compiling this book we aimed to bring together internationally diverse accounts of children's health and wellbeing as they relate to everyday geographies. The authors collectively take the reader on a journey to field sites in South and North America, Europe, Middle East and the South Pacific. However, any attempt to address the numerous threats to urban children's health and wellbeing in different social contexts and physical locations will be, and can only be, partial in scope.

We have organised the book into four parts: Neighbourhood environments; Home and away; Gardens, greens and nature; and Viewing wellbeing. The chapters in each section are diverse in subject matter but yet complementary in their concern for creating 'new' and 'better' stories of urban health and wellbeing.

Part 1 'Neighbourhood environments' considers how context and compositional aspects of different neighbourhoods determine children's healthy participation in urban life. Chapters in this section consider the ways symbolic and physical violence but also happiness shape neighbourhood experiences and feelings of belonging. Others investigate the impacts of environmental factors such as air pollution and the built environment on children's health and wellbeing. Akeson's chapter introduces the concept of 'miasma of occupation' to highlight the visible and invisible violence in Palestinian communities and its impact on children's physical, social and mental wellbeing. Daily life takes place against the backdrop of structural violence and inequalities of occupation fuelled by increasing settler violence. Children are growing up in a culture of conflict and distrust in highly politicised communities. Carver's chapter is set within the literature of active transport, independent mobility and stranger-danger. She argues that intervention at multiple scales is needed to counteract children's declining active transport in urban settings: recognition of physical and mental benefits of walking or cycling at the individual level; retrofitting built environments to meet the needs of pedestrians and cyclists; shifts in societal understandings of risk; and greater integration of new communication technologies to assist, promote and ease active transport and reduce associated risks and harms. Villaneuva and colleagues suggest expanding the research agenda on activity and urban neighbourhoods by focusing on relationships between the built environment and child development. They argue that a broader focus on children's cognitive and emotional development will engage decision makers more effectively and speak to the multi-faceted contexts in which children grow and develop. Lim and colleagues discuss the ways poor air quality impacts on mortality, morbidity and the quality of children's life. They argue that if the risks of air pollution are to be understood and effectively mitigated, children's exposure to air pollutants needs to be traced across time and space. Disabled children's rights to participate in everyday life through play are the subject of Wooley's chapter. She urges societies – following the social model of disability – to provide environments for disabled children that enable their participation in play activities. Freeman's chapter on happy cities concludes this section by introducing the reader to a raft of fun and happy events held in a New Zealand city offered to promote children's social, mental, and physical wellbeing. Freeman not only addresses the need to integrate child-friendly characteristics (good services and public places, safe environments) into city planning, but also critiques neoliberalism for undermining children's welfare and happiness. She challenges politicians, planners and child advocates to put children's happiness at the heart of all intervention and policy initiatives.

Part 2 'Home and away' sees authors focus on the meaning of 'home' for children's health and wellbeing. In these contributions, writers address the fluidity of the physical boundaries of children's home environments and question our current understanding and conceptualisations of children's health and wellbeing. Contributing to the independent mobility and environmental literacy debate, Nansen and colleagues focus on 'mobile device ecologies', indicating that physical boundaries of the home are expanded through parental gaze. Mobile devices literally become de facto companions that stretch spatial and temporal boundaries of children's everyday lives. Quiroz and Brooks discuss the creation of a new academic home environment catering for young black children in an elite neighbourhood and how it extended their academic potential as well as life chances more generally; however, in so doing they were alienated from their physical home and origins in a low socio-economic black neighbourhood. Their discussion examines the complexities of the geographies of educational opportunity. Vindrola-Patros and Brage's chapter examines how, in seeking cancer care in the capital of Argentina, rural children and their families' mobilities are shaped by economic, social, cultural and political factors. The emotional odyssey for finding the right treatment, as well as leaving family members and friends behind in their home towns, takes a financial and emotional toll on children and their families' physical and mental health and wellbeing. The authors signal more complex, dynamic and conflicting processes than have been imagined to date in the medical travel literature. Handelsman questions the traditional boundaries of childhood wellbeing by unpacking the cycles of violence in a shantytown community of Guayaquil, Ecuador. The overlapping cycles of motherhood and girlhood were fuelled by physical and structural violence, highlighting the fluid boundaries of childhood, adulthood and wellbeing.

Part 3 'Gardens, greens and nature' is focused on relationships between children's engagement with natural environments and their social, physical and mental wellbeing. Cushing and colleagues investigate access to healthy food choices in tropical Queensland, Australia. They combine discussions on the emerging urban food landscapes of community gardens and farmers' markets with debates on childhood obesity. These new urban food landscapes provide opportunities to educate about and entice children and their families into healthier eating practices through the enjoyment of seeing the produce growing and learning about the wellbeing of plants. Schäffer and Kraftl focus on the alternative educational spaces of forest kindergartens, forest schools and care farms in Germany and the UK. Their place- and class-based analysis elucidates strong inter-country differences in service rationale and provision adding a depth of understanding to nature education and its contribution to children's health and wellbeing. Similarly, Ergler and colleagues introduce the concept of 'wellbeing affordances' to expand the discussion of 'wild' nature spaces to children's everyday life in urban settings. By considering Auckland children's sense of place, place attachment, and spatial literacy they highlight how natural as well as sensory and aesthetically pleasing built environments can foster positive experiences and contribute to overall wellbeing of the individual child and the world at large.

Part 4 'Viewing wellbeing' expands and questions existing foci of childhood wellbeing through the lens of health promotion, Indigenous studies and political ecology. Masuda and colleagues argue for the need to be sensitive towards children's highly mobile lives in health promotion activities. In order to move beyond the 'fixed' representations of current wellbeing approaches, they suggest expanding current discussions of children's rights to the city, to seeing lives within, through and between places and socio-ecological contexts. They advocate shifting health promotion efforts from providing tools to 'fix' an individual's health and wellbeing to helping cities to become more equitable spaces per se. Thompson-Fawcett and Quigg's chapter addresses indigenous children's cultural wellbeing and identity in place. Referencing New Zealand's postcolonial context, they call for greater engagement with the language and spiritual histories of indigenous children in order to move towards an ethic of locatedness for addressing indigenous wellbeing more generally. Bartos and Wood introduce the concept of 'ecological wellbeing' set against the global change literature and also Bronfenbrenner's ecological systems theory to expand the research agenda of childhood wellbeing beyond the medical gaze. They argue that children's ecological wellbeing needs to be seen in the context of place, power and inter-relationality at the micro and macro scale. On the one hand, children shape and are shaped by the ecological wellbeing of the planet. On the other, we need to see their wellbeing bound by the structural contexts of 'globalised' societies. The authors conclude that a more encompassing understanding of wellbeing is required that pays attention to the diverse childhood ecologies literally and metaphorically.

References

Alderson, P. 2002. Young children's health care rights and consent. In: Franklin, B. (ed.) *The new handbook of children's rights: comparative policy and practice*. London: Routledge, pp. 155–76.

Andrews, G. J., Hall, E., Evans, B. and Colls, R. 2012. Moving beyond walkability: on the potential of health geography. *Social Science & Medicine*, 75(11), pp. 1925–32.

Andrews, H. 1985. The ecology of risk and the geography of intervention: from research to practice for the health and well-being of urban children. *Annals of the Association of American Geographers*, 75, pp. 370–82.

Atkinson, S., Fuller, S. and Painter, J. 2012. *Wellbeing and place*. Farnham: Ashgate.

Badland, H., Donovan, P., Mavoa, S., Oliver, M., Chaudhury, M. and Witten, K. 2015. Assessing neighbourhood destination access for children: development of the NDAI-C audit tool. *Environment and Planning B: Planning and Design*, 42(6), pp. 1148–60.

Barker, J. and Weller, S. 2003. 'Is it fun?' Developing children centred research methods. *International Journal of Sociology and Social Policy*, 23(1/2), pp. 33–58.

Beazley, H. 2003. The construction and protection of individual and collective identities by street children and youth in Indonesia. *Children, Youth and Environments*, 13(1). Available at www.colorado.edu/journals/cye/13_1/Vol13_1Articles/CYE_CurrentIssue_Article_ChildrenYouthIndonesia_Beazley.htm (accessed 22 February 2017).

Bingley, A. and Milligan, C. 2007. 'Sandplay, clay and sticks': multi-sensory research methods to explore the long-term mental health effects of childhood play experience. *Children's Geographies*, 5(3), pp. 283–96.

Blaut, J. M. 1987. Place perception in perspective. *Journal of Environmental Psychology*, 7(4), pp. 297–305.

Blaut, J. M. and Stea, D. 1971. Studies of geographic learning. *Annals of the Association of American Geographers*, 61(2) pp. 387–93.

Bond, E. 2014. *Childhood, mobile technologies and everyday experiences: changing technologies = changing childhoods?* London: Palgrave Macmillan.

Brown, T., McLafferty, S. L., Moon, G. and Ebooks, C. 2009. *A companion to health and medical geography*. Chichester: John Wiley & Sons.

Bunge, W. W. 1973. The geography. *Professional Geographer*, 25(4), pp. 331–7.

Bunge, W. W. 1977. The first years of the Detroit geographical expedition: personal report. In: Peet, R., (ed.) *Radical geography*. London: Methuen, pp. 31–9.

Cahill, C. 2007. The personal is political: developing new subjectivities through participatory action research. *Gender, Place & Culture*, 14(3), pp. 267–92.

Cardwell, F. S. and Crighton, E. J. 2014. Protecting our children: a scan of Canadian and international children's environmental health best practices. *Children, Youth and Environments*, 24(3), pp. 102–52.

Carroll, P., Asiasiga, L., Tav'ae, N. and Witten, K. 2013. Kids in the city: differing perceptions of one neighbourhood in Aotearoa/New Zealand. In: Coles, R. and Millman, Z. (eds.) *Landscape, wellbeing and environment*. London: Routledge, pp. 129–46.

Chawla, L. 1992. Childhood place attachment. In: Altman, I. and Low, S., (eds.) *Place attachment*. New York: Plenum Press, pp. 63–86.

Christensen, P. H. (ed.) 2000. *Research with children: perspectives and practices*. London: Falmer Press.

Clark, A. 2010. *Transforming children's spaces: children's and adults' participation in designing learning environments*. London: Routledge.

Collins, D. and Kearns, R. A. 2001. The safe journeys of an enterprising school: negotiating landscapes of opportunity and risk. *Health & Place*, 7(4), pp. 293–306.

Collins, D. and Kearns, R. A. 2005. Geographies of inequality: child pedestrian injury and walking school buses in Auckland, New Zealand. *Social Science & Medicine*, 60(1), pp. 61–69.

Cope, M. and Elwood, S. 2009. *Qualitative GIS: a mixed methods approach*. Thousand Oaks, CA: Sage.

Dahlgren, G. and Whitehead, M. 1991. *Policies and strategies to promote social equity in health*. Stockholm: Institute for Futures Studies.

de Carvalho, M. J. L. 2013. Children's perspectives on disorder and violence in urban neighbourhoods. *Childhood*, 20(1), pp. 98–114.

Derr, V. and Tarantini, E. 2016. 'Because we are all people': outcomes and reflections from young people's participation in the planning and design of child-friendly public spaces. *Local Environment*, 21(12), pp. 1–22.

Doherty, S. T., McKeever, P., Aslam, H., Stephens, L. and Yantzi, N. 2014. Use of GPS tracking to interactively explore disabled children's mobility and accessibility patterns. *Children, Youth and Environments*, 24(1), pp 1–24.

Donnelly, M. and Kilkelly, U. 2011. Child-friendly healthcare: delivering on the right to be heard. *Medical Law Review*, 19(1), pp. 27–54.

Ergler, C. R., Kearns, R. and Witten, K. 2013a. Managed childhoods: a social history of urban children's play. In: Higgins, N. and Freeman, C., (eds.) *Childhoods: growing up in Aotearoa New Zealand*. Dunedin, New Zealand: University of Otago Press, pp. 110–25.

Ergler, C. R., Kearns, R. A. and Witten, K. 2013b. Seasonal and locational variations in children's play: implications for wellbeing. *Social Science & Medicine*, 91, pp. 178–185.

Ergler, C. R., Kearns, R., Witten, K. and Porter, G. 2016. Digital methodologies and practices in children's geographies. *Children's Geographies*, 14(2), pp. 129–40.

Ergler, C. 2015. Beyond passive participation: from research on to research by children. In: Holt, L. and Evans, R., (eds.) *Geographies of children and young people. Methodological approaches*. Singapore: Springer Reference, pp. 1–19.

Ergler, C. 2016. Children's participation. In: Peters, M., (ed.) *Encyclopedia of educational philosophy and theory*. Singapore: Springer Reference, pp. 1–6.

Evans, B. 2006. 'Gluttony or sloth': critical geographies of bodies and morality in (anti) obesity policy. *Area*, 38(3), pp. 259–67.

Evans, B. and Colls, R. 2009. Measuring fatness, governing bodies: the spatialities of the body mass index (BMI) in anti-obesity politics. *Antipode*, 41(5), pp. 1051–83.

Fenton, N. E., Elliott, S. J. and Clarke, A. 2013. Tag, you're different: the interrupted spaces of children at risk of anaphylaxis. *Children's Geographies*, 11(3), pp. 281–97.

Fleuret, S. and Atkinson, S. 2007. Wellbeing, health and geography: a critical review and research agenda. *New Zealand Geographer*, 63, pp. 106–18.

Freeman, C. and Tranter, P. J. 2011. *Children and their urban environment: changing worlds*. London: Earthscan.

Freeman, C., Henderson, P. and Kettle, J. 1999. *Planning with children for better communities: the challenge to professionals*. Bristol: Policy Press.

Gallagher, M. 2008. 'Power is not an evil': rethinking power in participatory methods. *Children's Geographies*, 6(2), pp. 137–50.

Gesler, W. M. 1986. The uses of spatial analysis in medical geography: a review. *Social Science & Medicine*, 23(10), pp. 963–73.

Gesler, W. M. 1992. Therapeutic landscapes: medical issues in light of the new cultural geography. *Social Science & Medicine*, 34(7), pp. 735–46.

Gilliland, J., Rangel, C. Y., Healy, M. A., Tucker, P., Loebach, J. E., Hess, P. M., He, M., Irwin, J. and Wilk, P. 2012. Linking childhood obesity to the built environment: a multi-level analysis of home and school neighbourhood. *Canadian Journal of Public Health*, 103(9), pp. S15–21.

Glen, K. E., Thomas, H. M., Loebach, J. E., Gilliland, J. A. and Gobert, C. P. 2013. Fruit and vegetable consumption among children in a socioeconomically disadvantaged neighbourhood. *Canadian Journal of Dietetic Practice & Research*, 74(3), pp. 114–18.

Greenberg, M., Rosenberg, M., Phillips, D. and Schneider, D. 1990. Activism for medical geographers: American, British and Canadian viewpoints. *Social Science & Medicine*, 30(1), pp. 173–7.

Hart, R. 1979. *Children's experience of place*. New York: Irvington.

Hart, R., Satterthwaite, D., De La Barra, X. and Missair, A. 1999. *Cities for children: children's rights, poverty and urban management*. London: Earthscan.

Hayes, M. V. and Foster, L. 2002. *Too small to see, too big to ignore: child health and well-being in British Columbia*. Victoria, BC: Western Geographical Press.

Holloway, S. L. 2014. Changing children's geographies. *Children's Geographies*, 12(4), pp. 377–92.

Holloway, S. L. and Valentine, G. 2000. *Children's geographies: playing, living, learning*. New York: Routledge.

Holt, L. 2004. The 'voices' of children: de-centring empowering research relations. *Children's Geographies*, 2(1), pp. 13–27.

Holt, L. (ed.) 2011. *Geographies of children, youth and families: an international perspective*. London: Routledge.

Hörschelmann, K. and van Blerk, L. 2012. *Children, youth and the city*. London: Routledge.

Huo, X., Peng, L., Xu, X., Zheng, L., Qiu, B., Qi, Z., Zhang, B., Han, D. and Piao, Z. 2007. Elevated blood lead levels of children in Guiyu, an electronic waste recycling town in China. *Environmental Health Perspectives*, 115(7), pp. 1113–17.

James, A. and Prout, A. 1990. *Constructing and reconstructing childhood: contemporary issues in the sociological study of childhood*. London: Falmer Press.

James, A., Jenks, C. and Prout, A. 1998. *Theorizing childhood*. Cambridge: Polity Press.

Jones, O. 2001. 'Before the dark of reason': some ethical and epistemological considerations on the otherness of children. *Ethics, Place & Environment*, 4(2), pp. 173–8.

Kearns, R. A. 1997. Narrative and metaphor in health geographies. *Progress in Human Geography*, 21(2), pp. 269–77.

Kearns, R. A. and Andrews, G. J. 2010. Wellbeing. In: Smith, S., Pain, R., Marston, S. A. and Jones, J. P., (eds.) *Handbook of social geographies*. 3rd edn. London: SAGE, pp. 309–28.

Kearns, R. A. and Barnett, J. R. 2000. 'Happy meals' in the Starship Enterprise: interpreting a moral geography of health care consumption. *Health & Place*, 6(2), pp. 81–93.

Kearns, R. A. and Collins, D. C. A. 2000. New Zealand children's health camps: therapeutic landscapes meet the contract state. *Social Science & Medicine*, 51(7), pp. 1047–59.

Kearns, R. A. and Moon, G. 2002. From medical to health geography: novelty, place and theory after a decade of change. *Progress in Human Geography*, 26(5), pp. 605–25.

Knox, G. 1963. Detection of low intensity epidemicity application to cleft lip and palate. *British Journal of Preventive and Social Medicine*, 17, pp. 121–7.

Kraftl, P., Horton, J. and Tucker, F. 2007. Children, young people and built environments. *Built Environment*, 33(4), pp. 399–404.

Loebach, J. and Gilliland, J. 2010. Child-led tours to uncover children's perceptions and use of neighbourhood environments. *Children, Youth and Environments*, 20(1), pp. 52–90.

Loebach, J. E. and Gilliland, J. A. 2016. Free range kids? Using GPS-derived activity spaces to examine children's neighborhood activity and mobility. *Environment and Behavior*, 48(3), pp. 421–53.

Lynch, K. 1979. *Growing up in cities*. Cambridge, MA: MIT Press.

Maddison, R., Vander Hoorn, S., Jiang, Y., Ni Mhurchu, C., Exeter, D., Dorey, E., Bullen, C., Utter, J., Schaaf, D. and Turley, M. 2009. The environment and physical activity: the influence of psychosocial, perceived, and built environmental factors. *International Journal of Behavioral Nutrition and Physical Activity*, 6(1), pp. 19.

Mason, N. and Hanna, K. 2009. Undertaking child impact assessments in Aotearoa New Zealand local authorities: evidence, practice, ideas. Report for the Children's Commissioner. Wellington: Office of the Children's Commissioner.

Matthews, H. 2001. Power games and moral territories: ethical dilemmas when working with children and young people. *Ethics, Place & Environment*, 4(2), pp. 117–18.

Matthews, H. and Limb, M. 2003. Another white elephant? Youth councils as democratic structures. *Space and Polity*, 7(2), pp. 173–92.

Matthews, M. H. 1985. Young children's representations of the environment: a comparison of techniques. *Journal of Environmental Psychology*, 5(3), pp. 261–78.

Mavoa, S., Oliver, M., Witten, K. and Badland, H. 2011. Linking GPS and travel diary data using sequence alignment in a study of children's independent mobility. *International Journal of Health Geographics*, 10(1), pp. 64–73.

McKendrick, J. H. 2014. Geographies of children's wellbeing: in, of and for space. In: Ben-Arieh, A., Casas, F., Frone, I. and Korbin, J. E., (eds.) *Handbook of child wellbeing: theories, methods and policies in global perspective*. Dordrecht: Springer, pp. 279–300.

Mitchell, H., Kearns, R. A. and Collins, D. 2007. Nuances of neighbourhood: children's perceptions of the space between home and school in Auckland, New Zealand. *Geoforum*, 38(4), pp. 614–27.

Moore, R. C. 1986. *Childhood's domain: play and places in child development*. London: Croom Helm.

Neuwelt, P. and Kearns, R. A. 2006. Health benefits of walking school buses in Auckland, New Zealand: perception of children and adults. *Children, Youth and Environments*, 16(1), pp. 105–20.

NZ Herald, 2015a. Girls hit by car: 'drivers must take care'. 12 February 2015. Available at www.nzherald.co.nz/nz/news/article.cfm?c_id=1&objectid=11400754 (accessed 22 February 2017).

NZ Herald, 2015b. Too loud to play: Stonefields flying fox disabled. Available at www.nzherald.co.nz/nz/news/article.cfm?c_id=1&objectid=11388340 (accessed 22 February 2017).

Oliver, M., Mavoa, S., Badland, H. M., Carroll, P. A., Asiasiga, L., Tavae, N., Kearns, R. A. and Witten, K. 2014. What constitutes a 'trip'? Examining child journey attributes using GPS and self-report. *Children's Geographies*, 12(2), pp. 249–56.

Pain, R. 2006. Paranoid parenting? Rematerializing risk and fears for children. *Social & Cultural Geography*, 7(2), pp. 221–43.

Pearce, A., Kirk, C., Cummins, S., Collins, M., Elliman, D., Connolly, A. M. and Law, C. 2009. Gaining children's perspectives: a multiple method approach to explore environmental influences on healthy eating and physical activity. *Health & Place*, 15(2), pp. 614–21.

Pearce, J. and Witten, K., (eds.) 2010. *Geographies of obesity: environmental understandings of the obesity epidemic*. Aldershot: Ashgate.

Percy-Smith, B. and Thomas, N. 2010. *A handbook of children and young people's participation: perspectives from theory and practice*. New York: Routledge.

Philo, C. 2000. 'The corner-stones of my world': editorial introduction to special issue on spaces of childhood. *Childhood*, 7(3), pp. 243–56.

Piaget, J. and Inhelder, B. 1997. *The child's conception of space*. London: Routledge.

Punch, S. 2002. Youth transitions and interdependent adult–child relations in rural Bolivia. *Journal of Rural Studies*, 18(2), pp. 123–33.

Rainham, D., McDowell, I., Krewski, D. and Sawada, M. 2010. Conceptualizing the health-scape: contributions of time geography, location technologies and spatial ecology to place and health research. *Social Science & Medicine*, 70(5), pp. 668–76.

Rosenberg, M., 1998. Medical or health geography? Populations, peoples and places. *International Journal of Population Geography*, 4, pp. 211–26.

Schneider, D., Greenberg, M. and Lu, L. 1997. Early life experiences linked to diabetes mellitus: a study of African–American migration. *Journal National Medical Association*, 89(1), pp. 29–34.

Schwanen, T. and Atkinson, S. 2015. Geographies of wellbeing: an introduction. *The Geographical Journal*, 181(2), pp. 98–101.

Skinner, E. and Masuda, J. R. 2013. Right to a healthy city? Examining the relationship between urban space and health inequity by aboriginal youth artist–activists in Winnipeg. *Social Science & Medicine*, 91, pp. 210–18.

Smith, A. 2016. *Children's rights: towards social justice*. New York: Momentum Press.

Smith, N. R., Lewis, D. J., Fahy, A., Eldridge, S., Taylor, S. J., Moore, D. G., Clark, C., Stansfeld, S. A. and Cummins, S. 2015. Individual socio-demographic factors and perceptions of the environment as determinants of inequalities in adolescent physical and psychological health: the Olympic regeneration in East London (Oriel) study. *BMC Public Health*, 15(1), pp. 1–18.

Talbot, D. 2013. Early parenting and the urban experience: risk, community, play and embodiment in an east London neighbourhood. *Children's Geographies*, 11(2), pp. 230–42.

Taylor, N., Fitzgerald, R., Morag, T., Bajpai, A. and Graham, A. 2012. International models of child participation in family law proceedings following parental separation/divorce. *The International Journal of Children's Rights*, 20(4), pp. 645–73.

Taylor, Y. 2009. *Lesbian and gay parenting: securing social and educational capital*. Houndmills: Palgrave Macmillan.

UNICEF. 1995. *United Nations Convention on the Rights of the Child* [Online]. www.unicef. org/crc/ (accessed 20 March 2009).

Valentine, G. 2004. *Public space and the culture of childhood*. Aldershot: Ashgate.

Valentine, G., Butler, R. and Skelton, T. 2001. The ethical and methodological complexities of doing research with 'vulnerable' young people. *Ethics, Place & Environment*, 4(2), pp. 119–25.

van Blerk, L. and Kesby, M., (eds.) 2013. *Doing children's geographies: Methodological issues in research with young people*. London: Routledge.

Vine, M. M. and Elliott, S. J. 2013. Exploring the school nutrition policy environment in Canada using the Angelo framework. *Health Promotion Practice*, 15(3), pp. 331–9.

Ward, C. 1978. *The child in the city*. London: Architectural Press.

Winchester, H. P. M. 1991. The geography of children. *Area*, 23(4), pp. 357–60.

Witten, K., Kearns, R. A. and Carroll, P. 2015. Urban inclusion as wellbeing: exploring children's accounts of confronting diversity on inner city streets. *Social Science & Medicine*, 133, pp. 349–57.

Witten, K., Kearns, R. A., McCreanor, T. and Penney, L. 2009. Connecting place and the everyday practices of parenting: insights from Auckland, New Zealand. *Environment and Planning A*, 41, pp. 2893–910.

Yantzi, N. M., Young, N. L. and McKeever, P. 2010. The suitability of school playgrounds for physically disabled children. *Children's Geographies*, 8(1), pp. 65–78.

Part I

Neighbourhood environments

1   The miasma of occupation

The effects of seen and unseen violence on Palestinian children and families

Bree Akesson

Historically, miasma theory was used to explain diseases that were the result of 'bad air'. Although miasma theory has since been made obsolete by the development of germ theory in the late nineteenth century and other advances in public health, today the concept represents a constellation of unhealthy physical living conditions. In this chapter, miasma theory is applied to the context of Palestine where the geographical effects of occupation on children and families' wellbeing has been under-examined. Interviews with 18 Palestinian families (n = 149) revealed four factors contributing to the miasma of occupation in their neighbourhood communities: (1) increasing settler violence, (2) a childhood culture of conflict, (3) distrust between Palestinian neighbours and (4) the politicisation of neighbourhood communities. This chapter argues that these elements constitute the miasma of occupation, illustrating the damaging effects of structural inequalities on Palestinian children and families.

Contextualising Palestine

For the past century, the conflict between Israel and Palestine has continued to wreak havoc on the Palestinian and Israeli people and their land. Israel has occupied the Palestinian West Bank and Gaza Strip since 1967, marking the longest military occupation in modern history (Hajjar, 2005). Since then, violence has ebbed and flowed with tens of thousands – Israelis and Palestinians – injured or dead. Since the start of the first Palestinian *intifada* (uprising, in Arabic) in 1987, recurring hostilities between Palestinians and Israelis have left thousands of civilians dead and injured. These statistics have increased since 2008, which marked the start of three large rounds of hostilities focused in the Gaza Strip. Between January 2012 and December 2016, UNOCHA (2016) reported 2904 Palestinian deaths (70 per cent civilians) within the Gaza Strip, West Bank and Israel compared to 136 Israeli deaths (40 per cent civilians) within the same five-year time period.

Although the Israeli government maintains a public position of negotiating peace with Palestinians, it has also enacted an incremental process of Israeli territorialisation while simultaneously engaging in Palestinian *de*territorialisation (Yiftachel, 2006; Weizman, 2007). As a result, Palestinian geographical space has been shrinking under the weight of Israeli-driven policies, such as the control of the Palestinian movements, the continuing development of Israeli settlements in Palestinian land and violence that is a symptom of the occupation. These elements are described next.

Over the last decade, in the name of Israeli national security and supported by the international 'war on terror', Israel established a network of restrictions in Palestine, dividing Israelis and Palestinians and controlling Palestinian movement. These policies contribute to what Halper (2000) has identified as Israel's 'matrix of control'. The matrix of control includes elements such as the separation wall enclosing the West Bank, checkpoints that control Palestinian movement within the West Bank, the development of Israeli settlements on Palestinian land and complex policies of identification cards and travel permits (Weizman, 2007; Jones, 2012). The matrix of control cultivates artificial borders, deprives Palestinians of their freedom of movement and effectively suppresses the progress of human development within Palestine. Israel has faced broad international condemnation for these policies that carve off large segments of the West Bank, divide families and communities and separate Palestinians from their land. These policies also disrupt labour flows and basic commerce, eroding the productive capacity of the Palestinian economy.

For the most part, settlements are the main reason for the development of the devastating route of the separation wall and the ubiquitous checkpoints that make up the matrix of control. Settlements are organised residential, industrial and farming communities of Israeli civilians established on Palestinian land with the 'approval and direct or indirect support of the Israeli government' (UNOCHA, 2007: 13). However, settlement construction is considered to be a violation of international law (Defence for Children International (DCI), 2010). Since the beginning of the occupation in 1967, Israel has established some 200 settlements in the West Bank. Currently, over 350,000 Jewish settlers live in the area between the separation wall and the Green Line. An additional 300,000 settlers live in parts of East Jerusalem that Israel captured from Jordan in 1967 and later annexed in a move also considered illegal under international law (Sherwood, 2012; Rudoren and Ashkenas, 2015). Settlements have a profound effect on Palestinian life. Apart from the loss of Palestinian land taken for settlements, the seizure and/or destruction of property is an everyday occurrence in the lives of Palestinians, further contributing to the impoverishment of families and, as this chapter argues, the miasma of occupation.

As a place affected by political violence, Palestine is a landscape deeply inscribed with physical, emotional, cultural and political scars. To echo Tyner, violence in Palestine is both everywhere and nowhere: '. . . violence can be so pervasive, so prevalent, that we don't always 'see' violence' (2012: viii–ix). Fifty percent of children in Palestine are regularly exposed to various forms of violence in their homes, at school, on the streets or traveling between these places (DCI, 2008). Furthermore, DCI (2014) reports that each year approximately 500–700 Palestinian children, some as young as 12 years old, are detained, arrested and prosecuted by the Israeli military. Palestinian childhood is also characterised by violent experiences such as night raids, arrests of family members, home demolitions, construction of the wall around or through their communities, and personal assaults and injuries (Qouta *et al.*, 1995; Arafat and Boothby, 2003). Intra-Palestinian violence is also a threat to the wellbeing of Palestinian children. Hatred between Palestinian political groups, such as Fatah and Hamas, have resulted in hundreds of Palestinians being injured or killed (O'Callaghan *et al.*, 2009; B'Tselem, 2011a). Palestinian civilians have also been specifically targeted by Palestinian entities such as the Palestinian Authority, Fatah or Hamas on suspicion of collaboration with Israel. Collaboration has been broadly defined to include directly assisting Israel, agreeing with Israel's political positions, selling one's land to Israeli authorities, failing to participate in strikes, and marketing banned Israeli merchandise (B'Tselem, 2011b). Families of accused collaborators face extreme stigma and discrimination within their communities.

Violence permeates the unhealthy social and physical environment in which Palestinians live. However, this violence is not always directly related to the occupation. There are violent aspects of Palestinian society such as child abuse, family violence and gender-based violence that may be further exacerbated by factors such as unemployment, poverty and religious ideology. However, the occupation has ultimately created an environment where violence is commonplace and impacts on children's physical, social and emotional wellbeing. Ultimately, few Palestinian children are spared the osmotic effects of violence in their society (Baker and Shalhoub-Kevorkian, 1999; Thabet *et al.*, 2004; Shalhoub-Kevorkian, 2006).

Miasma theory

Prior to the development of germ theory in the late nineteenth century, diseases were thought to be the result of 'bad air'. The theory of 'miasma', a kind of polluting vapour that emerged from the accumulation of waste, prevailed as the main reason for disease (Tsuang *et al.*, 2011). At the end of the nineteenth century there was a dramatic change in understanding the cause of disease, shifting from miasma theory to infectious disease transmission (Tsuang *et al.*, 2011). At the same time, there was a shift in understanding that disease transmission was also related to social factors (Tsuang *et al.*, 2011). From this understanding, an approach conceptualised as the social determinants of health, which emphasises the broad social factors causing disease, replaced the previous focus on microorganisms (Tesh, 1995). Although miasma theory has been made obsolete by advances in public health, this chapter suggests that the concept of miasma can be revisioned under a social determinants of health framework by describing the unhealthy physical living conditions experienced by war-affected populations.

This chapter's understanding of miasma is related to the social determinants of health. Scholars have examined local patterns of distress and also the long-term health impact and psychosocial consequences of various forms of political violence (Pedersen, 2002), as well as the relationship between health inequalities and place (Bernard *et al.*, 2007). Summerfield (1998) notes that the effects of war cannot be separated from other factors such as structural poverty and injustice, and that these may 'undermine the social fabric no less effectively than the wars there have done' (Summerfield, 2000: 418). Social, political and economic injustice all contribute to a context of suffering, distress and disease (Pedersen, 2002). In other words, when trying to explain disease, distress, trauma and suffering in relation to violence, the issue of social inequality cannot be ignored as a critically important social determinant of health (Pedersen and Kienzler, 2014).

As an example of the relationship between modern interpretations of miasma theory and the social determinants of health, Epstein (2003) coined the term 'ghetto miasma' to describe disadvantaged neighbourhood communities with high morbidity and mortality rates in New York City.[1] People in these neighbourhood communities experienced stress from simply walking outside their homes:

> Poor parents, terrified that their kids will be killed on the street, tend to keep them inside, with the windows shut and the TV on, where they are constantly exposed to contaminants in indoor air, which some researchers believe can be as damaging as industrial pollution. . . . Mothers trying to protect their kids from crime may not realize they are putting their future health at risk.
>
> (Epstein, 2003: 6)

For the families in Epstein's report, the negative neighbourhood community created feelings of oppression and depression, especially because they felt as if they could not do anything to change their situation. Furthermore, the constant strain of oppression affected their sense of self-worth and dignity. With this thesis, Epstein suggested that constant stress and material deprivation are inseparable parts of the contemporary miasma of poverty and thereby important social determinants of health (Wilkins and Marmot, 2003). By concentrating the poor, any mysterious and poorly understood factor that makes people sick becomes clustered in these neighbourhood communities. Even though Epstein is describing spaces within New York City, her article provides strong parallels to neighbourhood communities in Palestine.

There is a large body of research that examines the effects of occupation on Palestinian children, and yet the impact of unhealthy living environments in Palestine has been under-researched. Even though community conditions are caused, to some extent, by violence, the priority has always been the *effects* of violence. However, studies of the poor and their living conditions indicate that there is a strong correlation between quality of neighbourhood community and wellbeing for families and children. Elements of neighbourhood communities that have been investigated in regards to children's development include residential instability, housing quality, noise, crowding, toxic exposure, quality of municipal services, recreational opportunities and the quality of educational and health facilities (Kopko, n.d.). Studies in the US show that people who live in disadvantaged neighbourhoods are more likely to have poor physical health than people who live in middle-class neighbourhoods, even when controlling for poverty (Diez-Roux, 2001; Cohen, Farley and Mason, 2003; Cohen, Mason, *et al.*, 2003). Furthermore, studies have found an association between chaotic environments and levels of psychological distress in children (Kopko, n.d.).

Methodology

This chapter represents part of a larger qualitative research project exploring the concept and meaning of place for children and families living in Palestine. In 2010, pilot interviews were conducted with Palestinian children, families and organisations.[2] Research continued in 2012, with the inclusion of families from various administrative regions of the occupied West Bank and annexed East Jerusalem, making a total of 18 participating families (149 individual participants). A minimum of three family members (parent; older child aged 9–18; and younger child aged 8 and under) were invited to take part in a collaborative interview focusing on their experiences with place. Interviews often included members of the larger extended family, or *hamula*, with some interviews including up to 12 *hamula* members. Family interviews – lasting between one and two hours – were conducted after each family member gave full and informed consent. Participants were guaranteed anonymity and assured that all information would remain confidential and used only for research purposes. Consistent with recommended and commonly used practices for conducting cross-language interviews (Knight *et al.*, 2009; Lopez *et al.*, 2008), two Palestinian research assistants (one in the West Bank and another in East Jerusalem) were hired and trained to translate during interviews and also to produce an interview transcript after the interview.

Data were collected using rapid ethnography (Millen, 2000; Handwerker, 2001; Mignone *et al.*, 2009). Like all forms of ethnography, rapid ethnography aims to gather a 'thick description' of participants' everyday lives and practices (Geertz, 1976), with participants selected for their unique cultural perspective and expertise (Handwerker, 2001). Rapid ethnography provides a reasonable understanding of the research participants and their practices given significant time pressures and limited time in the field (Millen, 2000). Using a rapid ethnographic approach, the researcher enters the cultural system with a specific data plan, identified informants and specific timelines because the researcher already has some familiarity with the issue and context (Handwerker, 2001). Based on my experience working in contexts of armed conflict and a field visit to Palestine prior to data collection, I chose a rapid ethnographic approach emphasising the core elements of a tight focus on children's relationships with place, the capture of rich data through the use of interactive techniques and multiple data gathering techniques to increase the likelihood of discovering new concepts and triangulating data (Millen, 2000). To further ground the data, ten interviews were conducted with key community informants who worked with Palestinian children and families.

A grounded theory approach (Charmaz, 2006) was used to analyse the data and was chosen for its systematic, yet flexible guidelines. Preliminary analytic memos helped to elaborate categories, specify their properties, define relationships between categories and identify gaps (Glaser and Strauss, 1967; Glaser, 1978; Strauss, 1987). Using Dedoose (www.dedoose.com), which is a web-based platform for qualitative data analysis, involved careful reading and annotation of the collated information to ascertain the meaning and significance of concepts that participants attributed to their experience living under occupation. Final themes were grouped around socio-spatial places that children and families interacted with in Palestine: home, school, neighbourhood community and nation-state. This chapter focuses on the place of neighbourhood community.

The miasma of occupation in Palestinian neighbourhood communities

Participants indicated that the occupation was characterised by four interrelated elements, which will be further described below: (1) violence from settlers, (2) a childhood culture of conflict, (3) distrust between Palestinians and (4) the politicisation of the Palestinian neighbourhood community.

Israeli settlers and the shrinking Palestinian space

In almost all field sites, families described the everyday violence from their Israeli settler neighbours. In Hebron, Abu-Jabar and his neighbour, Alia, explained how they 'don't get along with' their settler neighbours, painting a picture of increased isolation as more Palestinians move away and more settlers move in.[3] They told stories of their settler neighbours throwing soiled diapers and other trash into their yards. Abu-Majd described when six-year-old Mohamed was 'kidnapped' by Israeli settlers while playing on a wall dividing his home and a settler home.[4] According to Abu-Majd, after spending several hours with the settlers, Mohamed's family found him unconscious. Mohamed spent an additional five hours in the hospital being treated for shock. Abu-Majd reported the incident to the police, who refused to press charges because there were no photos or videos proving the incident took place. Mohamed and his family faced violence both seen and unseen. Mohamed was directly impacted by violence when taken by the settlers, and at the same time Abu-Majd experienced the unseen violence of structural inequality at the hands of the police.

Because of this seen and unseen violence, families described changes in their neighbourhood communities, most notably the shrinking of Palestinian space. Abu-Ali described the emptying of his neighbourhood community: 'Most of the neighbourhoods are empty. People are leaving from this area' and 'The majority is not here. It's almost empty. Most of the people left the neighbourhood'. The violence contributing to the shrinking neighbourhood communities was not only attributed to Israeli settlers and the military, but also to intra-Palestinian violence, which will be discussed in the next section.

A childhood culture of conflict

Many families noted that the violence was becoming ubiquitous within their neighbourhood communities, leading to the question of whether violence breeds further violence. The violence within these often small neighbourhood communities played out on multiple levels, including children acting this violence out. When I was walking to Abu-Ahmed's home in Hebron, a few Palestinian children on the nearby rooftops threw plastic pen caps at my head. I would feel an object hit my head and look up at the rooftops to see children laughing and trying to hide. The throwing escalated when the children began to fling metal spoons at me and my research assistant, after which we quickly moved inside the sanctuary of our host's home. In a recent article exploring the changing nature of play for Palestinian children in the West Bank, Rudoren (2013) reports that some children living in violent neighbourhood communities participate in role-playing games constructed around being arrested for throwing stones, which she named the 'West Bank culture of conflict'. The weapons of Israeli soldiers literally rule these children's lives, and it must be empowering therefore to feel as if one is holding that powerful weapon, even if it is make-believe – a symbolic challenge to the pervasive structural inequalities that these children live in. Families were aware of this 'culture of conflict' within their neighbourhood communities, with parents noting the bad influence of other children potentially influencing their own children, thereby contributing to the miasma of violence:

> Abu-Ahmed: There are good people and others are not. I don't mean bad people in ethics or morals. But maybe, as I told you, their children speak vulgar and bad words. Actually, I don't like my children to hang outside.

> Abu-Rachid: To be in the street is, um, doing nothing but bad things of course. What they will learn in the streets are bad things. Um, sometimes, the kids find that it is like play that they throw stones on soldiers. Imagine kids – seven [years old], six years old, nine years old. And . . . , according to [the Israelis], [stone throwing] is a dangerous thing to do. So, they started to arrest the children.

It is not just the bad behaviour of the children in the neighbourhood community, but also the potential consequences of this bad behaviour, such as arrest or retaliatory violence. This negativity and violence within the neighbourhood community led some parents to not allow their children to play outside, further limiting their place access (Akesson, 2014b) and contributing to the unhealthy miasma of occupation.

Distrust between Palestinians

Conflict existed not only between Palestinians and Israel settlers; some families also noted tensions between Palestinian neighbours. There were multiple reasons for this rooted in feelings of jealousy and a contrived set of criteria for who belongs to which place. For example, Umm-Omar noted that her neighbours are nice, but sometimes she does not get along with them 'because I am not originally from this village, and so I don't get the same respect that I would if I were from here'. In East Jerusalem, Umm-Ayoub faced similar judgment:

> I don't like it here. I don't like this house. And any time that I have a chance to go outside of the house, I will. The house we had in Hebron [was] very good, the house itself, but I don't like Hebron. And the extended family, you

know, the wives of the brothers of my husband are so mean. And, you know, family problems . . . They feel jealous.

Umm-Ayoub's daughter, Amina: . . .because she works and has a salary.

In these examples, the inequalities adults face trickle down and affect children. As a result, children may themselves face discrimination within the neighbourhood community.

In Hebron, Abu-Jabar and his neighbour Alia explain how their Palestinian neighbours had become more opportunistic rather than helpful:

Abu-Jabar: Sometimes, if you want to say good morning to your neighbour, it will cost you money or something else. Communication sometimes will cause you problems, so that's why we are not friendly with each other.

Interviewer: Why would somebody want money or why would it cause problems?

Alia: Their minds, their mentality, their thinking.

Abu-Jabar: People['s] psyches and minds differ from one person to another person. Everybody thinks of himself. There is a proverb that says, choose your friends before your road. Maybe because with these neighbours *(indicating Alia)*, our mentality and thought is the same, so we deal and talk together. Maybe, if I talk to somebody else, he will understand it wrong. Maybe he's opportunistic and abusive.

In contrast to a common conceptualisation of neighbourhood communities as positive environments, these examples indicate that they can be actually polarised and politicised.

Politicisation of the Palestinian neighbourhood-community

The history of the Oslo Accords epitomises the politicisation of neighbourhood communities. The first Palestinian *intifada* ended in 1993 with the signing of the Oslo Accords between the government of Israel and the Palestinian Liberation Organization (PLO), which in 1994 was formally established as the Palestinian Authority (PA). Although at the time they were hailed as a means to establish a permanent agreement for Palestinian control over the West Bank, after a period of five years, the Accords actually further fragmented the territory (Jones, 2012). According to Gordon (2008), the Oslo Accords outsourced the task of managing the Palestinian population to the PA, which was consequently responsible for the welfare of the maximum number of Palestinians in the least amount of land, whereas the least populated areas of the West Bank were left under direct Israeli control. The Accords resulted in the division of the West Bank, effectively turning Palestine into a 'Swiss cheese' state (Krouzman, 1999; Weizman, 2007). The Accords have become 'an empty shell', a typical example of a failed approach to reach a permanent solution between Israel and Palestine (Hitti, 2013) but, more importantly, families noted that the Accords were damaging to Palestinian communities. Zaki, the director of a community-based organisation that works with families in Bethlehem, was asked about the effects of Oslo. He replied: 'I am not [for Oslo], because it kills our souls. It kills everything in our life'. Zaki explained that before Oslo, there was solidarity among Palestinians, who collectively worked to improve the lives of their children, families and society. After Oslo, there was an influx of international organisations that brought in their own frameworks for assisting the Palestinian people. Zaki stated:

> Now, if I want to clean in front of my house, how much are you going to pay me? If I am going to participate in a demonstration, how much money am I going to receive? This is the problem after Oslo. Where is the youth? Where is the solidarity between people?

Zaki's comments are important when considering the significant role that international aid and development agencies play in shaping the lives and places of Palestinian children and their role in contributing to the miasma of occupation. Zaki explained that because of Oslo, Palestinians are more concerned with whether or not they will receive money from international organisations rather than fighting oppression, thereby leaving Palestinians powerless:

> We are waiting for the end of the month, whether there will be money for the salaries or not. This is what we care about. This is the thing we dream to reach, that when the end of the month is coming, I should have my salary. This is the result of Oslo. We are clients. We are clients waiting for the goods to come in order to eat it. That's it. There is no power for the society.

From Zaki's perspective, the culture of international aid as a result of the Oslo Accords created a dependent people rather than a Palestinian community based on solidarity and collective resistance. Abu-Jabar similarly explained that the neighbourhood community has changed as a result of the economic situation and the occupation and 'there is zero trust'. Families noted the prevalence of mistrust in the community because of potential spies,[5] and therefore families avoid interacting with one another.

Conclusion

The social determinants of health indicate that health and wellbeing is directly related to social factors such as stress, social exclusion, unemployment and opportunity for movement (Wilkins and Marmot, 2003). When applying this framework to the context of Palestine, it is easy to understand how chronic worry, an inability to protect one's family from violence, and an inability to change the toxic situation of occupation can contribute to structural inequality, grinding away at the human body and psyche and contributing to negative health and wellbeing. For example, the acknowledgement of elements of the miasma of occupation within partici-pants' disadvantaged neighbourhood communities was often not enough to make the families in this study want to leave this unhealthy place. In fact, despite the effects of the miasma of occupation, these families steadfastly remained *in place*. This response may be related to the fact that most families in the study did not have the financial means to leave – yet another form of structural inequality. In some cases, families noted that the neighbourhood was 'not good', but they still liked it because they felt 'comfortable', a contradiction of place explored in related research (Akesson, 2014a, 2014c). In fact, many of these same participants stated that they would defend their neighbourhood communities at all costs.

How can we address the miasma of occupation and thereby improve the health and wellbeing of Palestinian children and families living in these toxic neighbour-hood communities? Kaplan (1999) suggests that successful interventions must be informed by a socio-historical approach to place. Kaplan continues to emphasise the importance of an interdisciplinary approach, 'venturing out into an area of interven-tion where the success is uncertain but the payoff could be dramatic' (1999: 118). This research represents one step towards better understanding the experiences of Palestinian children and families affected by the miasma of occupation.

To further understand the intersection between place and structural inequalities, additional contexts should be also examined. For example, although refugee camps are intended to serve as a short-term solution for displaced populations, they often eventually become semi-permanent cities (Ledwith, 2014) with inhabitants facing further challenges, such as sub-standard sanitation, lack of livelihoods and vio-lence. Several studies have found that the most important risk factor challenging the wellbeing of children was prolonged settlement in a refugee camp (Thabet and Vostanis, 1998; Khamis, 2005; Giacaman *et al.*, 2007). Findings from this chapter indicate that it is valuable to consider how structural inequalities contribute to long-term psychosocial and health outcomes for these marginalised populations in order to ameliorate the negative effects and improve health and wellbeing.

To conclude, the interviews with Palestinian children and families revealed four specific factors contributing to the miasma of occupation in their neigh-bourhood communities. First, settlers remain a major threat to the wellbeing of Palestinian children and families contributing to the shrinking of Palestinian space. Second, the occupation itself has added to a childhood culture of conflict whereby violence becomes a normal element of Palestinian childhood. Third, distrust between Palestinian neighbours has led to a disintegration of community support. Finally, the politicisation of Palestinian neighbourhood communities has led to divided neighbourhood communities with distrust between neighbours. These four elements make up the miasma of occupation for Palestinian children and families and illustrate how damaging the multi-faceted aspects of structural inequality can be to the social determinants of health. Returning to Tyner's under-standing of violence as being 'both everywhere and nowhere' (2012: viii), the miasma of occupation is characterised by violence and is exacerbated by structural inequalities. It is a noxious and ubiquitous presence – both seen and unseen – that infiltrates Palestinian neighbourhood communities.

Notes

1 I use the term neighbourhood community in this research to convey both the physical and social elements of place. For a more detailed definition, see Akesson, 2014a: 140.
2 Human subjects' approval was obtained through the appropriate Research Ethics Board prior to commencement of the study. All names and identifying details about the participants have been changed to ensure anonymity.
3 In Arabic culture, after the birth of the first son, parents are given the honorary title of *abu* (father of) and *umm* (mother of) followed by the name of the first-born son. I have used these identifiers throughout this chapter to indicate fathers and mothers from the index family.
4 Palestinians sometimes use the term 'kidnapped' to describe when one is arrested.
5 Israel has been known to offer some Palestinians money or favour in exchange for their collaboration. Palestinians caught collaborating with Israel are often shunned or killed. Families of Palestinians who have been caught spying for Israel suffer discrimination and isolation in their communities (Ayyoub, 2013).

References

Akesson, B. 2014a. *Contradictions in place: everyday geographies of Palestinian children and families living under occupation*. PhD thesis. McGill University, Montréal, QC.
Akesson, B. 2014b. Arrested in place: Palestinian children and families at the border. In: S. Spyrou, and M. Christou, eds. *Children and borders: studies in childhood and youth*. Basingstoke: Palgrave Macmillan, pp. 81–98.
Akesson, B. 2014c. Castle and cage: meanings of home for Palestinian children and families. *Global Social Welfare*, 1, pp. 81–95.
Arafat, C. and Boothby, N. 2003. *A psychosocial assessment of Palestinian children*. US Agency for International Development, West Bank and Gaza.
Ayyoub, A. 2013. *Families of Palestinian spies for Israel face stigma in Gaza*. Al-Monitor.
Baker, A.M. and Shalhoub-Kevorkian, N. 1999. Effects of political and military traumas on children: the Palestinian case. *Clinical Psychological Review*, 19, pp. 935–50.
Bernard, P., Charafeddine, R., Frohlich, K.L., Daniel, M., Kestens, Y. and Potvin, L. 2007. Health inequalities and place: a theoretical conception of neighbourhood. *Social Science & Medicine*, 65, pp. 1839–852.
B'Tselem. 2011a. *Severe human rights violations in inter-Palestinian clashes*. B'Tselem, Jerusalem.
B'Tselem. 2011b. *Harm to Palestinians suspected of collaborating with Israel*. B'Tselem, Jerusalem.
Charmaz, K. 2006. *Constructing grounded theory: a practical guide through qualitative analysis*. London: SAGE.
Cohen, D., Farley, T. and Mason, K. 2003. Why is poverty unhealthy? Social and physical mediators. *Social Science & Medicine*, 57, pp. 1631–41.
Cohen, D., Mason, K., Bedimo, A., Scribner, R., Basolo, V. and Farley, T. 2003. Neighborhood physical conditions and health. *American Journal of Public Health*, 93, pp. 467–71.
Defence for Children International (DCI). 2008. *Under attack: settler violence against Palestinian children in the Occupied Territory*. Jerusalem: DCI.
Defence for Children International (DCI). 2010. *Under attack: settler violence against Palestinian children in the occupied Palestinian territory* [update]. Jerusalem: DCI.
Defence for Children International (DCI). 2014. *Solitary confinement for Palestinian children in Israeli military detention*. Jerusalem: DCI.

Diez-Roux, A.V. 2001. Investigating neighborhood and area effects on health. *American Journal of Public Health*, 91, pp. 1783–9.

Epstein, H. 2003. Ghetto miasma: enough to make you sick? *New York Times Magazine*, 12 October.

Geertz, C. 1976. Thick description: towards an interpretive theory of culture. In C. Geertz, ed. *The interpretation of cultures*. New York: Basic Books, pp. 3–30.

Giacaman, R., Abu-Rmeileh, N., Husseini, A., Saab, H. and Boyce, W. 2007. Humiliation: the invisible trauma of war for Palestinian youth. *Public Health*, 121, 563–71; discussion 572–7.

Glaser, B.G. 1978. *Theoretical sensitivity: advances in the methodology of grounded theory*. Mill Valley, CA: Sociology Press.

Glaser, B.G. and Strauss, A.L. 1967. *The discovery of grounded theory: strategies for qualitative research*. Chicago, IL: Aldine.

Gordon, N. 2008. *Israel's occupation*. Berkeley, CA: University of California Press.

Hajjar, L. 2005. *Courting conflict: The Israeli military court system in the West Bank and Gaza*. London: University of California Press.

Halper, J. 2000. The 94 percent solution: a matrix of control. *Middle East Report*, 216, pp. 14–19.

Handwerker, W.P. 2001. *Quick ethnography*. Walnut Creek, CA: AltaMira Press.

Hitti, N. 2013. Oslo Accords offer lessons after twenty-year 'interim'. *Al-Monitor*, 10 September.

Jones, R. 2012. *Border walls: security and the war on terror in the United States, India, and Israel*. London: Zed Books.

Kaplan, G.A. 1999. What is the role of the social environment in understanding inequalities in health? *Annals of the New York Academy of Sciences*, 896, pp. 116–19.

Khamis, V. 2005. Post-traumatic stress disorder among school age Palestinian children. *Child Abuse & Neglect*, 29, pp. 81–95.

Knight, G.P., Roosa, M.W. and Umaña-Taylor, A.J. 2009. *Studying ethnic minority and economically disadvantaged populations: methodological challenges and best practices*. Washington, DC: American Psychological Association.

Kopko, K. n.d. *The effects of the physical environment on children's development: insights for parents, teachers, and educators featuring research by Dr Gary Evans, Departments of Human Development and Design and Environmental Analysis, Cornell University*. Ithaca, NY: Cornell University College of Human Ecology.

Krouzman, R. 1999. Twenty-first century Palestine: toward a 'Swiss-cheese' state? *Middle East Report*, 29.

Ledwith, A. 2014. *Zaatari: the instant city*. Boston, MA: Affordable Housing Institute.

Lopez, G.I., Figueroa, M., Connor, S.E. and Maliski, S.L. 2008. Translation barriers in conducting qualitative research with Spanish speakers. *Qualitative Health Research*, 18, pp. 1729–37.

Mignone, J., Hiremath, G.M., Sabnis, V., Laxmi, J., Halli, S., O'Neil, J., Ramesh, B.M., Blanchard, J. and Moses, S. 2009. Use of rapid ethnographic methodology to develop a village-level rapid assessment tool predictive of HIV infection in rural India. *International Journal of Qualitative Methods*, 8, pp. 68–83.

Millen, D.R. 2000. Rapid ethnography: time deepening strategies for HCI field research. *Proceedings of the 3rd Conference on Designing Interactive Systems: Processes, Practices, Methods, and Techniques, DIS '00*. New York: ACM, pp. 280–6.

O'Callaghan, S., Jaspars, S. and Pavanello, S. 2009. *Losing ground: protection and livelihoods in the occupied Palestinian territory. Humanitarian Policy Group (HPG) Working Paper*. London: Overseas Development Institute.

Pedersen, D. 2002. Political violence, ethnic conflict, and contemporary wars: broad implications for health and social well-being. *Social Science & Medicine*, 55, pp. 175–90.

Pedersen, D. and Kienzler, H. 2014. Mental health and illness outcomes in civilian populations exposed to armed conflict and war. In: S.O. Okpaku, ed. *Essentials of global mental health*. New York: Cambridge University Press, pp. 307–15.

Qouta, S., Punamäki, R.-L. and El-Sarraj, E. 1995. The impact of the peace treaty on psychological well-being: a follow-up study of Palestinian children. *Child Abuse & Neglect*, 19, pp. 1197–208.

Rudoren, J. 2013. In a West Bank culture of conflict, boys wield the weapon at hand. *New York Times*, 4 August.

Rudoren, J. and Ashkenas, J. 2015. Netanyahu and the settlements. *New York Times*, 12 March.

Shalhoub-Kevorkian, N. 2006. Negotiating the present, historicizing the future: Palestinian children speak about the Israeli separation wall. *American Behavioral Scientist*, 49, pp. 1101–24.

Sherwood, H. 2012. Population of Jewish settlements in West Bank up 15,000 in a year. *Guardian*, 26 July.

Strauss, A.L. 1987. *Qualitative analysis for social scientists*. Cambridge: Cambridge University Press.

Summerfield, D. 1998. The social experience of war and some issues for the humanitarian field. In: P.J. Bracken, and C. Petty, eds. *Rethinking the trauma of war*. London: Free Association Books/Save the Children, pp. 9–37.

Summerfield, D. 2000. Childhood, war, refugeedom and 'trauma': three core questions for mental health professionals. *Transcultural Psychiatry*, 37, pp. 417–33.

Tesh, S.N. 1995. Miasma and 'social factors' in disease causality: lessons from the nineteenth century. *Journal of Health Politics Policy and Law*, 20, pp. 1001–24.

Thabet, A.A.M. and Vostanis, P. 1998. Social adversities and anxiety disorders in the Gaza Strip. *Archives of Disease in Childhood*, 78, pp. 439–42.

Thabet, A.A.M., Abed, Y. and Vostanis, P. 2004. Comorbidity of PTSD and depression among refugee children during war conflict. *Journal of Child Psychology & Psychiatry*, 45, pp. 533–42.

Tsuang, M.T., Tohen, M. and Jones, P. 2011. *Textbook of psychiatric epidemiology*. Chichester: John Wiley & Sons.

Tyner, J.A. 2012. *Space, place, and violence: violence and the embodied geographies of race, sex, and gender*. New York, NY: Routledge.

United Nations Office for the Coordination of Humanitarian Affairs (UNOCHA). 2007. *The humanitarian impact on Palestinians of Israeli settlements and other infrastructure in the West Bank*. Jerusalem: UNOCHA.

United Nations Office for the Coordination of Humanitarian Affairs (UNOCHA). 2016. *Monthly figures*. Jerusalem: UNOCHA and Occupied Palestinian Territory. Available at https://www.ochaopt.org/content/monthly-figures# (accessed 28 February 2017).

Weizman, E. 2007. *Hollow land: Israel's architecture of occupation*. London: Verso.

Wilkins, R. and Marmot, M., eds. 2003. *The social determinants of health: the solid facts*, 2nd edn. Denmark: World Health Organization.

Yiftachel, O. 2006. *Ethnocracy*. Philadelphia, PA: Pennsylvania University Press.

2 Children's active transport and independent mobility in the urban environment

Alison Carver

Children's active transport (e.g. walking/cycling to local destinations) and independent mobility (i.e. their freedom to move around their neighbourhood or similar without adult accompaniment; Hillman *et al.*, 1990) are key behaviours that contribute to children's health and wellbeing. However, compared with previous generations, children in developed nations engage less in active transport and have lower levels of independent mobility (Hillman *et al.*, 1990; Valentine, 1997; Shaw *et al.*, 2013). This chapter describes how these behaviours are beneficial to children's physical and mental health and why declines in these behaviours are of public health concern. Linking with the field of health geography (Brown *et al.*, 2009) and guided by ecological models of health behaviours (Sallis and Owen, 1997), I examine social and physical environmental predictors of children's unsupervised active travel and outdoor play in urban neighbourhoods. Interventions and strategies aimed at promoting walking and cycling to school and local places are described, along with implications for urban planning and policy.

How childhood has changed: declines in active transport and independent mobility

The nature of childhood in the developed world has changed markedly over recent decades (Hörschelmann and van Blerk, 2012; Higgins and Freeman, 2013). Although contemporary children are afforded greater material benefits (such as access to organised sports programs and individual bedrooms that provide indoor play-space with electronic entertainment media) (Karsten, 2005; Hillman, 2006), their freedom to walk, cycle and play in their neighbourhood without adult supervision has diminished (Hillman *et al.*, 1990; Shaw *et al.*, 2013). During the 1950s and 1960s almost all children were considered as 'outdoor children' who played alfresco (Karsten, 2005). Although 'outdoor children' still exist, two further categories have evolved: 'indoor children' who spend much of their leisure time indoors, engaging in sedentary behaviours such as television viewing and playing electronic games; and 'the backseat generation' who are regularly driven by car to structured, adult-led sports programmes or music/dancing lessons (Karsten, 2005). Declines in active transport to school are reported in the USA (McDonald *et al.*, 2011), the UK (Hillman *et al.*, 1990; Pooley *et al.*, 2005) and Australia (Salmon *et al.*, 2005). Children's independent mobility has declined (Hillman *et al.*, 1990; Shaw *et al.*, 2013) and it is less common to see children playing ball games or mingling on local streets (Hillman, 2006).

Health benefits of active transport and independent mobility

Declines in children's active transport and independent mobility are of public health concern because both behaviours are potential sources of habitual physical activity (Faulkner *et al.*, 2009; Schoeppe *et al.*, 2013) for which health benefits include reduced risk of cardiovascular disease, type-2 diabetes, obesity and some cancers (Trost, 2005). Further benefits for children include increased bone mineral density (Bailey and Martin, 1994; Strong *et al.*, 2005; Trost, 2005) and enhanced mental wellbeing (Calfas and Taylor, 1994; Trost, 2005). However, many children are insufficiently physically active. Having identified physical inactivity as the fourth highest factor for global mortality, the World Health Organization (2010) recommends school-aged children accrue at least 60 minutes of moderate-to-vigorous intensity physical activity (e.g. walking briskly or cycling) each day. Australian guidelines also recommend that less than two hours per day are spent in sedentary behaviours, particularly during daylight hours (Australian Government Department of Health, 2014). Children's independent mobility confers further benefits by promoting their social, cognitive and emotional development (Kyttä, 2004). Independent mobility in particular promotes children's bonding with peers (Prezza *et al.*, 2001), a sense of community and less concern about crime during adolescence (Prezza and Pacilli, 2007).

Physical activity accrued by walking, cycling, or outdoor play in the neighbourhood is particularly important in areas of low socio-economic status (SES). In developed countries, such as Australia, engagement in leisure-time physical activity has been shown to follow a social gradient, whereby the socio-economically disadvantaged are less likely to participate in habitual physical activity and are more likely to have poorer health outcomes than those who are advantaged (Australian Bureau of Statistics, 2013). In particular, children who reside in low SES neighbourhoods (Ziviani *et al.*, 2007) tend to engage in lower physical activity levels than those in high SES neighbourhoods, and this is also true for children whose families are disadvantaged rather than advantaged (Lee and Cubbin, 2002; Woodfield *et al.*, 2002). Structured commercially-available sports programs, such as organised team sports, tend to be particularly less accessible to low SES children for whom the cost (e.g. fees, equipment) may be prohibitive (Ziviani *et al.*, 2007). However, walking, cycling and outdoor play are free or low-cost activities and present vital sources of habitual physical activity for low SES children. Overall, the neighbourhood is generally an accessible setting for independently mobile children because parents are not required to provide transport to nearby venues.

Social and physical environmental predictors of children's active transport and independent mobility

To understand why children engage in low levels of active transport and lack independent mobility it is important to identify predictors of these behaviours. Ecological models of health behaviours (Sallis and Owen, 1997) posit that there are multiple layers of influence of intrapersonal, social and physical environmental variables centred on the individual. It is well-established that intrapersonal factors such as age and sex are associated with active transport and independent mobility, with boys rather than girls and older rather than younger children tending to walk or cycle more for transport and also to have greater autonomy (Hart, 1979; Hillman *et al.*, 1990; Valentine, 1997; Prezza *et al.* 2001). Social and physical environmental factors that may influence these behaviours among urban children in their neighbourhood will be discussed next.

Seminal research on children's independent mobility was conducted during 1971–1990 in England by Hillman and colleagues (Hillman *et al.*, 1990). They identified that nearly four times as many children were chauffeured to school in 1990 compared to 1971, and that 'traffic danger' was the main reason for parental restriction of children's active transport (Hillman *et al.*, 1990: 24). At that time, the Department of Transport was attributing reduced rates of child pedestrian injury and fatality to improved road safety. However, Hillman and colleagues argued that road environments had not become safer, but rather that children were being kept indoors due to parental concerns about road safety and, thus, were not exposed to the road environment (Hillman *et al.*, 1990).

A literature review by Carver *et al.* (2008) identified that key reasons for parental restriction of their children's active transport and outdoor play in their neighbourhood were concerns about road safety and potential harm from strangers (or 'stranger danger'). These concerns and their role as barriers to children's active transport and independent mobility remain pervasive among more recent study findings (Mammen, 2012; Carver, Timperio *et al.*, 2013; O'Connor and Brown, 2013). However, further reasons for restricting these behaviours have been identified in the context of the school journey. For example, because family routines have become complex, the school journey is often made by car and incorporated into parental commutes to work (Mitra, 2013) or linked with shopping trips or organised extra-curricular activities (Pooley *et al.*, 2005). Even on short walkable trips car travel is considered convenient, time-saving and socially acceptable (Mackett, 2003; Lang *et al.*, 2011; Trapp *et al.*, 2011; Mitra, 2013).

A consistent barrier to active transport on the school journey is distance between home and school (Davison *et al*, 2008; Yeung *et al.*, 2008; Wong *et al.*, 2011; Carver *et al.*, 2012; Trapp *et al.*, 2011; Su *et al.*, 2013; Oliver *et al.*, 2014). Although it has been suggested that urban planners should locate schools close to homes (Mammen *et al.*, 2012; Su *et al.*, 2013), the solution is less simplistic in situations where parents have freedom of choice of school, even within state-provided education systems if facilities or standards vary (Carver, Watson *et al.*, 2013). Hillman (2006) reported that children in England were travelling further to school following the introduction of the Education Reform Act 1988 (HMSO, 1988), which allowed greater choice of schools.

Neighbourhood 'walkability' appears frequently in literature on active transport and this concept has been measured both subjectively and objectively. For example, Panter *et al.* (2010) examined perceptions of the walking/cycling infrastructure, neighbourhood safety and proximity to destinations, whilst Saelens *et al.* (2003) measured 'walkability' using objective measures of street connectivity, footpaths and pedestrian crossings. Evidence suggests that neighbourhood walkability is related to children's active transport (Panter *et al.*, 2010; Giles-Corti *et al.*, 2011; Christiansen *et al.*, 2014). Furthermore, a New Zealand study (Oliver *et al.*, 2014) examined neighbourhood self-selection regarding walkability as a predictor of active transport to school, and reported that children residing in 'low-walkable' neighbourhoods were less likely to walk or cycle to school if their parents had a preference for 'high-walkable' neighbourhoods. Conversely, children residing in 'low-walkable neighbourhoods' and whose parents preferred to live in 'low-walkable' neighbourhoods, as well as all children residing in 'high-walkable' neighbourhoods, were more likely to use active transport to school (Oliver *et al.*, 2014). This study highlights the complexity of social and environmental influences on children's everyday behaviours and also the crucial role played by parental values and practices.

Interventions and strategies to promote children's active transport and independent mobility

Interventions that aim to encourage children's active transport and independent mobility may take diverse approaches. These include social interventions, changes to the design of the built environment (e.g. the introduction of infrastructure for pedestrians and cyclists) and community/societal interventions. Examples of these will now be discussed along with the potential for communications technology to play a role in promoting children's independent mobility.

One broad-reaching intervention that has been implemented in many developed nations (Tudor-Locke *et al.*, 2001; Collins and Kearns, 2005; Hillman, 2006; VicHealth, 2007) to promote children's active transport on the school journey is the Walking School Bus (WSB), whereby children are picked up (and later dropped off) at 'bus stops' close to home and walk together to/from school along a defined route, led by an adult 'driver' and followed by an adult 'conductor' (VicHealth, 2007). WSBs were initially conducted in the UK and New Zealand in the 1990s (Collins and Kearns, 2005) and pilot-testing of this program took place in Victoria, Australia between 2001 and 2002 (VicHealth, 2007). VicHealth (2007) reported that as well as increasing physical activity levels, participating children benefited from opportunities for social interaction and there was less traffic congestion around school gates. Although similar benefits were demonstrated in New Zealand (Kearns *et al.*, 2003), a lack of parent volunteers who are required to lead such programs (Tudor-Locke *et al.*, 2001) resulted in lower prevalence of WSBs in socio-economically disadvantaged areas (Collins and Kearns, 2005). In addition to sustainability issues from lack of volunteer staff, there has been criticism of the WSB program by some transportation researchers (Tranter and Malone, 2003). Hillman (2006) in particular suggests that because of such programs, some parents may feel neglectful if their child does not always have adult accompaniment when outdoors.

A systematic review of interventions that aimed to promote active transport to school (Chillon *et al.*, 2011) demonstrated that those which engaged schools, parents and local communities *concurrently* showed potential in increasing rates of active transport. That review of 14 separate interventions identified various intervention components that included installation of traffic signals, improvement of footpaths, mapping of safe routes and WSB (Chillon *et al.*, 2011). Although heterogeneity in design and methodology of those interventions precluded identification of the most successful strategies, evidence suggested that specifically targeting active transport was more effective than promoting broader health behaviours simultaneously (Chillon *et al.*, 2011). However, in this emerging research field further interventions are required with robust experimental designs, valid and reliable measures, and appropriate data analyses (Chillon *et al.*, 2011; Stewart *et al.*, 2014).

When promoting active transport to school, a one-size-fits-all approach should be avoided in areas where children travel non-walkable distances. Initiatives to promote active transport during longer journeys might include combining walking or cycling with public transport (Carver, Veitch *et al.*, 2014). 'Park and stride' facilities near schools encourage children to walk the remaining distance after being driven to a designated area and are inversely associated with sedentary behaviour (van Sluijs *et al.*, 2011). However, in areas where children live far from school, it may be more worthwhile to enable opportunities for active travel to extra-curricular activities or destinations close by to their home (Smith *et al.*, 2012; Carver, Timperio *et al.*, 2013).

The need to promote children's active transport and independent mobility in the neighbourhood is further emphasised by Hillman (2006) who argues that school journeys represent only a fraction of all journeys made. Short journeys to local destinations present potential intervention points for active transport instead of habitual car travel. For example, an Australian study identified that parents of primary schoolchildren were making an average of three to four trips per week driving their children to destinations that parents considered to be within walking distance of home (Carver, Timperio *et al.*, 2013).

Comprehensive literature reviews have identified that aspects of the built environment associated with children's physical activity and active transport in their neighbourhoods include low traffic volume and speed, access to sports and recreational facilities, infrastructure for pedestrians and cyclists, and mixed land use with co-location of residences, commercial entities and civic facilities in order that there are appropriate walkable/cycleable destinations within reach of home (Pont *et al.*, 2009; Ding *et al.*, 2011). The importance of appropriate infrastructure is emphasised by evidence of higher rates of cycling in areas with dedicated cycling infrastructure. An Australian study reported that children were more likely to cycle regularly in neighbourhoods where bike paths were prevalent (Carver, Timperio *et al.*, 2014). Furthermore, higher rates of cycling in the Netherlands, Denmark and Germany, compared with other developed nations, are attributed to the cycling infrastructure being separated from roads in high-traffic areas, combined with traffic calming measures (e.g. speed bumps) on residential streets (Pucher and Buehler, 2008).

As described earlier, concerns about road safety and 'stranger danger' are major reasons for parental restriction of children's walking, cycling and outdoor play in their neighbourhoods (Carver *et al.*, 2008). Although examination of road accident statistics has identified children as vulnerable road users (UNICEF, 2001), less evidence exists of strangers randomly attacking or abducting children, who are more likely to be assaulted by a family member or acquaintance (Finkelhor and Ormrod, 2000; Shutt *et al.*, 2004). Despite this, a report by the Criminology Research Council (1998: 9) in Australia describes a 'risk-victimization paradox' whereby parents are overly anxious about their children's safety and exaggerate the risk of harm from strangers. Parents may justify their concerns by citing publicised cases of children being harmed by strangers or by describing 'hypothetical worst-case scenarios' (Tulloch, 2004: 15). Road safety concerns may be reduced by introducing traffic calming measures and such infrastructure is associated with children's physical activity (Carver *et al.*, 2010). However, little research has explored strategies to mitigate concern about stranger danger in relation to children's active transport and independent mobility.

An Australian study (Foster *et al.*, 2014) reported that parental fear of strangers was associated with lower odds of children being independently mobile, regardless of the supportiveness of their social and built environments. In particular, the notion that neighbours would look out for children who live nearby failed to reduce the influence of parental fear on their children's independent mobility (Foster *et al.*, 2014). Conversely, in other countries, such as Germany, this collective responsibility where adults help or exert authority over unaccompanied children in public is considered to be associated with children's independent mobility (Hillman *et al.*, 1990; Tranter and Pawson, 2001). As suggested by Foster *et al.* (2014) interventions aimed at increasing children's independent mobility may be ineffective unless there is greater understanding of factors that influence parental fears of harm from strangers. O'Connor and Brown (2013) reported that parents are torn between wishing to promote their child's autonomy and fearing what might happen. To address this, O'Connor and Brown (2013) suggest that interventions promote the benefits of independent active transport rather than dwell on whether related fears are irrational.

Mass media reports of child abductions are often cited as influencing parents' perceptions of stranger danger (O'Connor and Brown, 2013, Foster *et al.*, 2014); however, little research has examined how media reporting impacts perceived risk of harm from strangers, and how this affects parents' rules and behaviours regarding their children's active transport and independent mobility. Furthermore, as communications technology develops, it is recognised that localised reports regarding crime and community safety may be transmitted via social media, text messages and emails (O'Connor and Brown, 2013). This type of reporting may serve as both a barrier to, and a facilitator of, children's active transport and independent mobility and warrants closer investigation.

The use of communications technology in promoting active transport and independent mobility should be explored, particularly the role of social media to promote the health benefits of these behaviours. In addition, mobile phones may be

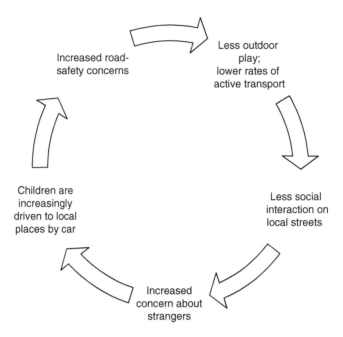

Figure 2.1 How concerns about road safety and potential harm from strangers are related.
Source: based on Mullan, 2003.

useful by providing parents with opportunities for remote surveillance (Fotel and Thomsen, 2004). For example, parents may allow their children to travel independently on the condition that they call their parent on arrival (Fotel and Thomsen, 2004, Nansen *et al.*, this volume). Apps are being designed for smartphones to promote child safety, for example the 'Help Me' app assists children in distress and was launched in Australia following the high-profile case of the abduction and murder of Daniel Morecombe, a 13-year-old schoolboy (Daniel Morcombe Foundation, 2012). Although children carrying mobile phones may give parents greater confidence to allow their child to venture out alone (Carver *et al.*, 2012), mobile phones do not necessarily promote independent mobility. In fact, a Norwegian study (Hjorthol, 2008) found that mobile phone ownership promoted greater car use because parents responded to their child's most typical text message 'Come and pick me up'.

Conclusion

It may be more difficult to address less tangible concerns about stranger danger rather than road safety concerns. However, by improving road safety for pedestrians and cyclists it may be possible to address stranger danger concerns, which Mullan (2003) suggests are closely linked with road safety concerns (see Figure 2.1).

Mullan (2003) proposes that due to road safety concerns there is less outdoor play, walking and cycling on local streets, and subsequently less social interaction. If parents have limited contact with neighbours, they may worry about potential harm from strangers and consider it necessary to drive their children rather than allow them to walk or cycle to local destinations. This increases traffic volume and heightens road safety concerns. Mullan (2003) describes the interrelationship of these concerns and behaviours as a 'downward spiral of fear'. Improvements in the safety of the road environment may therefore potentially encourage more walking, cycling and outdoor play in the local neighbourhood and, indirectly, reduce concern about strangers.

There is broader evidence that neighbourhoods with physically active residents are socially cohesive. For example, an Irish study demonstrates that adults who resided in a neighbourhood that was conducive to walking were more likely to interact with neighbours, to consider themselves as members of the local community and to trust others (Leyden, 2003). Kawachi and Berkman (2000: 175) describe social cohesion as comprising a pair of societal attributes: 'the absence of latent social conflict' and 'the presence of strong social bonds'. The former attribute refers to social harmony despite diversity of financial status, religious beliefs or political inclinations, while the latter includes mutual trust and assistance (i.e. 'social capital') and infrastructure, as well as channels for conflict resolution (Kawachi and Berkman, 2000: 175). A further construct related to this point is 'collective efficacy', which 'refers to shared expectations and mutual engagement by residents in local social control' (Sampson, 2003: 138). The building of collective efficacy, which is inversely associated with local violence, may be an important strategy to increase neighbourhood safety for children and the broader community (Sampson, 2003) and also to encourage their active transport and overall physical activity levels.

All things considered, factors that may influence children's active transport and independent mobility within their neighbourhoods are complex. It is suggested that urban planners include an appropriate mix of land use in residential areas in order that commercial entities, schools and sports facilities are located within walking or cycling distance of homes, and that local road networks are supportive of walking and cycling (e.g. with traffic-calming measures, footpaths, bike paths). This may be challenging if urban planners operate within political contexts (e.g. in New Zealand) where larger schools are preferred for reasons related to economic efficiency (Lewis *et al.*, 2004). However, further research is required to identify innovative ways of reducing parental concerns about safety and promoting the benefits of these behaviours for children. It appears that 'one-size-fits all' solutions may not be possible and targeted approaches may be more effective. Interventions that provide safe road environments and render neighbourhoods conducive to active transport and outdoor play will not alleviate all safety concerns. However, building collective efficacy within neighbourhoods by encouraging greater social interaction between residents of all ages, as well as increasing children's outdoor play, cycling and pedestrian activity on local streets, may play a preliminary role in lessening the perceived risk of harm from strangers.

References

Australian Government Department of Health. 2014. *Australia's Physical Activity and Sedentary Behavior Guidelines 2014*. Available at www.health.gov.au/internet/main/publishing.nsf/Content/health-pubhlth-strateg-phys-act-guidelines (accessed 27 February 2017).

Australian Bureau of Statistics. 2013. *Australian Health Survey: Physical Activity, 2011–12.* Available at www.abs.gov.au/ausstats/abs@.nsf/Lookup/462FBA87B642FCA4CA257BAC0015F3CE (accessed 19 February 2015).

Bailey, D.A. and Martin, A.D. 1994. Physical activity and skeletal health in adolescents. *Pediatric Exercise Science*, 6(4), pp. 330–47.

Brown, T., McLafferty, S.L. and Moon, G. 2009. *A companion to health and medical geography*. Chichester: Wiley.

Calfas, K. and Taylor, W. 1994. Effects of physical activity on psychological variables in adolescents. *Pediatric Exercise Science*, 6, pp. 406–23.

Carver, A., Timperio, A. and Crawford, D. 2008. Playing it safe: the influence of neighbourhood safety on children's physical activity – a review. *Health & Place*, 14(2), pp. 217–27.

Carver, A., Timperio, A. and Crawford, D. 2012. Young and free? A study of independent mobility among urban and rural dwelling Australian children. *Journal of Science and Medicine in Sport*, 15(6), pp. 505–10.

Carver, A., Timperio, A. and Crawford, D. 2013. Parental chauffeurs: what drives their transport choice? *Journal of Transport Geography*, 26(1), pp. 72–7.

Carver, A., Timperio, A. and Crawford, D. 2014. Bicycles gathering dust rather than raising dust – prevalence and predictors of cycling among Australian schoolchildren. *Journal of Science and Medicine in Sport*, 18(5), pp. 540–4.

Carver, A., Timperio, A., Hesketh, K. and Crawford, D. 2010. Are safety-related features of the road environment associated with smaller declines in physical activity among youth? *Journal of Urban Health*, 87(1), pp. 29–43.

Carver, A., Veitch, J., Sahlqvist, S., Crawford, D. and Hume, C. 2014. Active transport, independent mobility and territorial range among children residing in disadvantaged areas. *Journal of Transport & Health*, 1(4), pp. 267–73.

Carver, A., Watson, B., Shaw, B. and Hillman, M. 2013. A comparison study of children's independent mobility in England and Australia. *Children's Geographies*, 11(4), pp. 461–475.

Chillon, P., Evenson, K., Vaughn, A. and Ward, D. 2011. A systematic review of interventions for promoting active transportation to *school. International Journal Behavioural Nutrition & Physical Activity*, 8(1), p. 10.

Christiansen, L.B., Toftager, M., Schipperijn, J., Ersbøll, A.K., Giles-Corti, B. and Troelsen, J. 2014. School site walkability and active school transport – association, mediation and moderation. *Journal of Transport Geography*, 34(1), pp. 7–15.

Collins, D.C.A. and Kearns, R.A. 2005. Geographies of inequality: child pedestrian injury and walking school buses in Auckland, New Zealand. *Social Science & Medicine*, 60(1), pp. 61–9.

Criminology Research Council. 1998. *Fear of crime*. Available at http://crg.aic.gov.au/reports/1998-foc1.pdf (accessed 27 February 2017).

Daniel Morcombe Foundation Inc. 2012. *Help Me app*. Available at www.danielmorcombe.com.au/app.html (accessed 1 September 2014).

Davison, K., Werder, J. and Lawson, C. 2008. Children's active commuting to school: current knowledge and future directions. *Preventing Chronic Disease*, 5(3), p. A100.

Ding, D., Sallis, J.F., Kerr, J., Lee, S. and Rosenberg, D.E. 2011. Neighborhood environment and physical activity among youth. *American Journal of Preventive Medicine*, 41(4), pp. 442–55.

Faulkner, G.E.J., Buliung, R.N., Flora, P.K. and Fusco, C. 2009. Active school transport, physical activity levels and body weight of children and youth: a systematic review. *Preventive Medicine*, 48(1), pp. 3–8.

Finkelhor, D. and Ormrod, R. 2000. *Characteristics of crimes against juveniles*. Washington, DC: Office of Juvenile Justice and Delinquency Prevention.

Foster, S., Villanueva, K., Wood, L., Christian, H. and Giles-Corti, B. 2014. The impact of parents' fear of strangers and perceptions of informal social control on children's independent mobility. *Health & Place*, 26, pp. 60–8.

Fotel, T. and Thomson, U. 2004. The surveillance of children's mobility. *Surveillance & Society*, 1, pp. 535–54.

Giles-Corti, B., Wood, G., Pikora, T., Learnihan, V., Bulsara, M., van Niel, K., Timperio, A., McCormack, G. and Villanueva, K. 2011. School site and the potential to walk to school: the impact of street connectivity and traffic exposure in school neighborhoods. *Health & Place*, 17(2), pp. 545–50.

Hart, R.,1979. *Children's experience of place*. New York: Irvington.

Her Majesty's Stationery Office (HMSO). 1988. *The Education Reform Act 1988*. London: HMSO.

Higgins, N. and Freeman, C. (eds.) 2013. *Childhoods: growing up in Aotearoa New Zealand*. Dunedin: University of Otago Press.

Hillman, M. 2006. Children's rights and adults' wrongs. *Children's Geographies*, 4(1), pp. 61–7.

Hillman, M., Adams, J. and Whitelegg, J. 1990. *One false move . . .: a study of children's independent mobility*. London: PSI Publishing.

Hjorthol, R. 2008. The mobile phone as a tool in family life: impact on planning of everyday activities and car use. *Transport Reviews*, 28(3), pp. 303–20.

Hörschelmann, K. and van Blerk, L. 2012. *Children, youth and the city*. London: Routledge.

Karsten, L. 2005. It all used to be better? Different generations on continuity and change in urban children's daily use of space. *Children's Geographies*, 3(3), pp. 275–90.

Kawachi, I. and Berkman, L.F. (eds.) 2000. *Social epidemiology*. New York: Oxford University Press, pp. 174–90.

Kearns, R.A., Collins, D.C.A. and Neuwelt, P.M. 2003. The walking school bus: extending children's geographies? *Area*, 35, pp. 285–92.

Kyttä, M. 2004. The extent of children's independent mobility and the number of actualized affordances as criteria for child-friendly environments. *Journal of Environmental Psychology*, 24(2), pp. 179–98.

Lang, D., Collins, D. and Kearns, R. 2011. Understanding modal choice for the trip to school. *Journal of Transport Geography*, 19(4), pp. 509–14.

Lee, R.E. and Cubbin, C. 2002. Neighborhood context and youth cardiovascular health behaviors. *American Journal of Public Health*, 92(3), pp. 428–36.

Lewis, N. 2004. Embedding the reforms in New Zealand schooling: after neo-liberalism? *GeoJournal*, 59, pp. 149–60.

Leyden, K.M. 2003. Social capital and the built environment: the importance of walkable neighborhoods. *American Journal of Public Health*, 93(9), pp. 1546–51.

Mackett, R. 2003. Why do people use their cars for short trips? *Transportation*, 30, pp. 329–49.

Mammen, G., Faulkner, G., Buliung, R. and Lay, J. 2012. Understanding the drive to escort: a cross-sectional analysis examining parental attitudes towards children's school travel and independent mobility. *BMC Public Health*, 12(1), p. 862.

McDonald, N., Brown, A., Marchetti, L. and Pedroso, M. 2011. US school travel, 2009 an assessment of trends. *American Journal Preventative Medicine*, 41(2), pp. 146–51.

Mitra, R. 2013. Independent mobility and mode choice for school transportation: a review and framework for future research. *Transport Reviews*, 33(1), pp. 21–43.

Mullan, E. 2003. Do you think that your local area is a good place for young people to grow up? The effects of traffic and car parking on young people's views. *Health & Place*, 9(4), pp. 351–60.

O'Connor, J. and Brown, A. 2013. A qualitative study of 'fear' as a regulator of children's independent physical activity in the suburbs. *Health & Place*, 24, pp. 157–64.

Oliver, M., Badland, H., Mavoa, S., Witten, K., Kearns, R., Ellaway, A., Hinckson, E., Mackay, L. and Schluter, P. 2014. Environmental and socio-demographic associates of children's active transport to school: a cross-sectional investigation from the URBAN Study. *International Journal of Behavioral Nutrition and Physical Activity*, 11(1), p. 70.

Panter, J.R., Jones, A.P., van Sluijs, E.M.F. and Griffin, S.J. 2010. Attitudes, social support and environmental perceptions as predictors of active commuting behaviour in school children. *Journal of Epidemiology and Community Health*, 64(1), pp. 41–8.

Pont, K., Ziviani, J., Wadley, D., Bennett, S. and Abbott, R. 2009. Environmental correlates of children's active transportation: a systematic literature review. *Health & Place*, 15(3), pp. 849–62.

Pooley, C.G., Turnbull, J. and Adams, M. 2005. The journey to school in Britain since the 1940s: continuity and change. *Area*, 37(1), pp. 43–53.

Prezza, M. and Pacilli, M.G. 2007. Current fear of crime, sense of community, and loneliness in Italian adolescents: the role of autonomous mobility and play during childhood. *Journal of Community Psychology*, 35(2), pp. 151–70.

Prezza, M., Pilloni, S., Morabito, C., Sersante, C., Alparone, F.R. and Giuliani, M.V. 2001. The influence of psychosocial and environmental factors on children's independent mobility and relationship to peer frequentation. *Journal of Community & Applied Social Psychology*, 11, pp. 435–50.

Pucher, J. and Buehler, R. 2008. Making cycling irresistible: lessons from The Netherlands, Denmark and Germany. *Transport Reviews*, 28(4), pp. 495–528.

Saelens, B.E., Sallis, J.F., Black, J.B. and Chen, D. 2003. Neighborhood-based differences in Physical Activity: an environment scale evaluation. *American Journal of Public Health*, 93(9), pp. 1552–8.

Sallis, J. and Owen, N. 1997. Ecological models. In: K. Glanz, F. Lewis and B. Rimer (eds.), *Health behaviour and health education: theory, research, and practice*. 2nd edn. San Francisco, CA: Jossey-Bass, pp. 403–24.

Salmon, J., Timperio, A., Cleland, V. and Venn, A. 2005. Trends in children's physical activity and weight status in high and low socio-economic status areas of Melbourne, Victoria, 1985–2001. *Australian and New Zealand Journal of Public Health*, 29(4) pp. 337–42.

Sampson, R.J. 2003. Neighborhood-level context and health: lessons from sociology. In: I. Kawachi and L. Berkman (eds.), *Neighborhoods and health*. New York: Oxford University Press, pp. 132–46.

Schoeppe, S., Duncan, M.J., Badland, H., Oliver, M. and Curtis, C. 2013. Associations of children's independent mobility and active travel with physical activity, sedentary behaviour and weight status: a systematic review. *Journal of Science and Medicine in Sport*, 16(4), pp. 312–19.

Shaw, B., Watson, B., Frauendienst, B., Redecker, A., Jones, T. and Hillman, M. 2013. *Children's independent mobility: a comparative study in England and Germany (1971–2010)*. London: Policy Studies Institute.

Shutt, E.J., Miller, M.J., Schreck, C.J. and Brown, N. 2004. Reconsidering the leading myths of stranger child abduction. *Criminal Justice Studies: A Critical Journal of Crime, Law and Society*, 17(1), pp. 127–34.

Smith, L., Sahlqvist, S., Ogilvie, D., Jones, A., Griffin, S.J. and van Sluijs, E. 2012. Is active travel to non-school destinations associated with physical activity in primary school children? *Preventive Medicine*, 54(3–4), pp. 224–8.

Stewart, O., Moudon, A.V. and Claybrooke, C. 2014. Multistate evaluation of safe routes to school programs. *American Journal of Health Promotion*, 28(3), pp. S89–96.

Strong, W.B.M.D., Malina, R.M.P., Blimkie, C.J.R.P., Daniels, S.R.M.D.P., Dishman, R.K.P., Gutin, B.P., Hergenroeder, A.C.M.D., Must, A.P., Nixon, P.A.P., Pivarnik, J.M.P., Rowland, T.M.D., Trost, S.P. and Trudeau, F.P. 2005. Evidence-based physical activity for school-age youth. *Journal of Pediatrics*, 146(6), pp. 732–37.

Su, J., Jerrett, M., McConnell, R., Berhane, K., Dunton, G., Shankardass, K., Reynolds, K., Chang, R. and Wolch, J. 2013. Factors influencing whether children walk to school. *Health & Place*, 22, pp. 153–61.

Tranter, P. and Malone, K. 2003. *Out of bounds: insights from children to support a cultural shift towards sustainable and child-friendly cities*. Paper presented to State of Australian Cities National Conference, Parramatta, NSW, 3–5 December.

Tranter, P. and Pawson, E. 2001. Children's access to local environments: a case-study of Christchurch, New Zealand. *Local Environment*, 6(1), pp. 27–48.

Trapp, G.S.A., Giles-Corti, B., Christian, H.E., Bulsara, M., Timperio, A.F., McCormack, G.R. and Villaneuva, K.P. 2011. Increasing children's physical activity. *Health Education & Behavior*, 39(2), pp. 172–82.

Trost, S. 2005. *Discussion paper for the development of recommendations for children's and youth's participation in health promoting physical activity*. Canberra: Australian Government Department of Health and Ageing.

Tudor-Locke, C., Ainsworth, B.E. and Popkin, B.M. 2001. Active commuting to school: an overlooked source of children's physical activity? *Sports Medicine*, 31(5), pp. 309–13.

Tulloch, M. 2004. Parental fear of crime: a discursive analysis. *Journal of Sociology*, 40(4), pp. 362–77.

UNICEF, 2001. *A league table of child deaths by injury in rich nations, No. 2*. Florence: UNICEF Innocenti Research Centre.

Valentine, G. 1997. 'My son's a bit dizzy.' 'My wife's a bit soft': gender, children, and cultures of parenting. *Gender, Place and Culture*, 4(1), pp. 37–62.

van Sluijs, E., Jones, N., Jones, A., Sharp, S., Harrison, F. and Griffin, S. 2011. School-level correlates of physical activity intensity in 10-year-old children. *International Journal Pediatric Obesity*, 6(2–2), e574–81.

VicHealth. 2007. *Walking School Bus*. Available at www.vichealth.vic.gov.au/wsb (accessed 26 February 2015).

Wong, B.-M., Faulkner, G. and Buliung, R. 2011. GIS measured environmental correlates of active school transport: a systematic review of 14 studies. *International Journal Behaviour Nutrition & Physical Activity*, 8, p. 39.

Woodfield, L., Duncan, M., Al-Nakeeb, Y., Nevill, A. and Jenkins, C. 2002. Sex, ethnic and socio-economic differences in children's physical activity. *Paediatric Exercise Science*, 14(3), pp. 277–85.

World Health Organization (WHO). 2010. *Global recommendations on physical activity for health*. Geneva: WHO.

Yeung, J., Wearing, S. and Hills, A.P. 2008. Child transport practices and perceived barriers in active commuting to school. *Transportation Research Part A: Policy and Practice*, 42(6), pp. 895–900.

Ziviani, J., Wadley, D., Ward, H., Macdonald, D., Jenkins, D. and Rodger, S. 2007. A place to play: socioeconomic and spatial factors in children's physical activity. *Australian Occupational Therapy Journal*, 55(1), pp. 2–11.

3 How does the neighbourhood built environment influence child development?

Karen Villanueva, Hannah Badland and Melody Oliver

Neighbourhoods provide important exposures and resources that can influence the parents' capacity to raise their children and promote healthy child development. Although impacts of environmental exposure on children's behaviours have been explored by epidemiologists and children's geographers, child development research has largely ignored neighbourhood contexts, particularly that of the built (physical) neighbourhood environment. There is growing evidence that the built environment influences children's physical health through its impact on physical activity behaviours, such as active play, walking and cycling, and independent mobility. For example, children living in more walkable, pedestrian-friendly neighbourhoods are more likely to be physically active, including walking and cycling to destinations, when compared with those living in less walkable neighbourhoods (see Chapter 2). Children's interaction with, and exposure to, their neighbourhood largely occurs through active travel and play. These experiences, in turn, impact their mental, social, emotional, cognitive and physical development.

Given that neighbourhoods are among the most common settings where children spend time outside of home and school, there is substantial potential for the neighbourhood built environment to facilitate or hinder positive child development. In this chapter, we discuss which features of the neighbourhood built environment may be conducive for healthy child development, and highlight gaps in the empirical evidence to support this relationship. The imperatives for this chapter are twofold: first, the global interest in creating liveable, sustainable and equitable neighbourhoods; and second, the centrality of local experience in the 'child-friendly cities' movement. We acknowledge at the outset that the work reviewed has largely been conducted about rather than with children. However, we see development of a coherent picture of the links between neighbourhood built environments and child development as an important prerequisite to the development of a more child-inclusive research agenda.

The importance of growing up in health-promoting environments

Ensuring healthy foundations are laid early in life is crucial to children's wellbeing (Zubrick *et al.*, 2010) and, as a consequence, the health of young children is recognised as a global public health priority (UNICEF, 2013). Children's physical development, social competence, emotional maturity, language and cognitive development, and general knowledge and communication skills are interrelated aspects of healthy early development (Farrar *et al.*, 2007). Poorer outcomes in childhood can negatively influence physical, social and economic outcomes later in life. Evidence indicates developmental inequities emerging in early childhood are maintained (and often widen) into adulthood, being expressed as higher rates of mortality and physical, social and cognitive morbidity across the social gradient (Hertzman, 2010; Nicholson *et al.*, 2012).

Health inequities, including *between* and *within* cities and neighbourhoods, arise from disparities in social, economic, political, cultural and environmental factors (Commission on Social Determinants of Health, 2008). As the World Health Organization states, the social determinants of health (SDH) are 'the conditions in which people are born, grow, live, work, and age . . .' (Commission on Social Determinants of Health, 2008). Public health approaches seeking to reduce inequalities recognise the importance of the SDH as part of the causal pathway to child development outcomes, (see Chapter 1 on neighbourhood violence as an SDH for Palestinian children's wellbeing) and as a potential opportunity for intervention. Various socio-ecological models have been readily accepted in child developmental theory and have recognised neighbourhood contexts as potential influences (Bronfenbrenner, 1979; Sampson *et al.*, 2002); however, most development research has focused on proximal social environments, specifically family and schools (Leventhal and Brooks-Gunn, 2000).

Although research into the factors influencing child development is not new, the true burden of child health inequalities is unknown because the effects of the neighbourhood – the physical and social place in which children are born, grow and age – have been largely overlooked (see Chapter 16).

Why explore the nexus between *built* environment and child development?

Where we live can substantially influence our health and social behaviours, impacting through the SDH pathways (Sallis *et al.*, 2012). Physical neighbourhood examples of the SDH encompass the broader inanimate neighbourhood context and can include features such as access to affordable housing, facilities and services; provision of cycling and public transport infrastructure; local employment and education opportunities; appropriate social infrastructure (e.g. schools, child care, local shops); and healthy food options. The way in which neighbourhoods and cities are designed has received increasing attention over recent years, given more than half of the global population live in cities, with projections of exponential population growth by 2050 (United Nations, 2010).

Significant global health and planning agencies (e.g. United Nations, World Health Organization) accordingly advocate for building sustainable, 'liveable' and equitable neighbourhoods and cities to support our growing population. Such environments can promote healthier lifestyles, enhance quality of life and reduce inequity across the life course (Australian Government Department of Infrastructure and Transport, 2013). Within this context, the focus on vulnerable subpopulations, such as children, is increasing. National and global movements on creating 'child-friendly cities' (Gleeson and Sipe, 2006) respond to this agenda by reflecting children's rights in policies and programmes within urban settings. Such agendas emphasise the importance of children's involvement in shaping the places and spaces in which they grow (Freeman and Tranter, 2011). Despite policy interest, there is substantial evidence suggesting that children's access to and movement through their neighbourhood environments is declining in many countries (Fyhri *et al.*, 2011; Carver *et al.*, 2013). The urban built environment, as one of many factors that may contribute to the decline, is examined in this chapter.

Connecting urban design with child health and development can bring together multiple key players, including paediatricians, public health practitioners, urban designers and planners as well as families and children to advocate and consider more community-level approaches and place-based initiatives that promote children's healthy development. Exploring how the built environment impacts child development is an important first step. In its simplest form, it is hypothesised that this relationship is through children's physical and social interaction with their neighbourhood (see Figure 3.1).

Neighbourhood physical interaction

Children's interaction with, and exposure to, their local built environment largely occurs through physical activity behaviours. For very young children (e.g. 0–5 years), physical interaction with their surroundings is largely through active play, i.e. 'unstructured outdoor physical activity that takes place in a child's free time' (Brockman *et al.*, 2011: 1). Active play is vital for young children's development and wellbeing; not only does it promote physical activity, but it also encourages social interaction, supports creativity and facilitates problem-solving (Wood and Martin, 2010, Brockman *et al.*, 2011). As children age, they become more mobile; their environmental interactions and experiences contribute to their social, emotional, cognitive and physical development (Badland and Oliver, 2012). Physical activity behaviours such as walking or cycling, independent mobility (i.e. travel without adult supervision), sport and exercise are important for normal healthy child development (Armstrong, 1993). Physically active children are more likely to have better motor, spatial and cognitive skills, and enhanced mental health and social wellbeing, which positively impact on outcomes in later life (Castonguay and Jutras, 2009). Diverse built environment attributes are associated with children's physical activity (Sallis *et al.*, 2000). By way of example, children living in more walkable neighbourhoods (characterised by well-connected streets with safe crossing points, footpaths, mix of local destinations present and low traffic volumes and speeds) are more likely to be physically active and walk and cycle to destinations when compared to those living in less walkable communities (Kerr *et al.*, 2006). This evidence base is fairly well developed for associations between built environment attributes and selected children's health behaviours; however, the same cannot

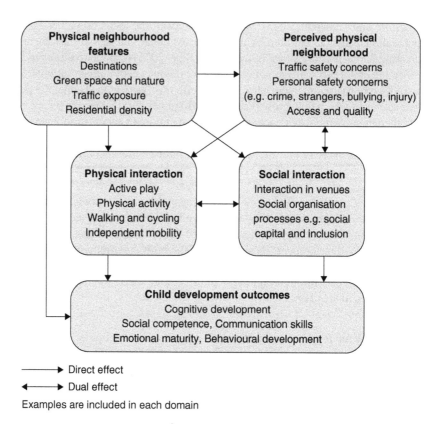

Figure 3.1 Pathways in which the neighbourhood built environment may affect child development.

Source: Villanueva, K., Badland, H., Kvalsvig, A., O'Connor, M., Christian, H., Woolcock, G., Giles-Corti, B. and Goldfeld, S. (unpublished).

be said for associations between built environment attributes and child development outcomes.

Neighbourhood social interaction

Being outdoors in the public realm can offer a 'third place' in which informal (chance meetings) and formal (planned meetings) interactions can take place (Jacobs, 1961). Neighbourhood design has a role in helping to facilitate children's play and social interaction (Wood *et al.*, 2010) through creating opportunities for these behaviours to occur (Carroll *et al.*, 2015). For example, more walkable neighbourhoods have been found to increase neighbourhood social interaction because people are more likely to walk, cycle and linger in these environments (Leyden, 2003). One implication of neighbourhood social interaction is better social capital (Wood and Giles-Corti, 2008). In particular, the presence of more children 'out and about' in neighbourhoods may help create stronger local communities by encouraging feelings of safety, trust and communality (Tranter and Whitelegg, 1994; Wood *et al.*, 2013). As children are more likely to be outdoors if they (and/or their parents) feel safe (Carver *et al.*, 2008), they are more likely to access community resources and have opportunities for meeting and developing relationships with other local children and neighbours (Prezza and Pacilli, 2007). For children, the benefits of higher social capital may include improved general health, adoption of healthy behaviours and healthy development (Badland and Oliver, 2012).

Importantly, although research suggests a strong theoretical platform for expecting the neighbourhood *built environment* to influence child development, such attributes have been somewhat neglected in child development research (Herrington and Studtmann, 1998).

Based on the evidence available, four built environment features appear to be associated with child development: neighbourhood destinations; green spaces and nature; traffic; and housing and density. Each is discussed in the following sections.

The role of the built environment on child development

Neighbourhood destinations

Destinations such as schools, recreational settings, child care, and health and social service providers provide children with opportunities to be social, observe, interact and explore their neighbourhood (Giles-Corti *et al.*, 2009).

Child development studies have explored the influence of destinations and services in light of inequalities within and between neighbourhoods (Jencks and Mayer, 1990; Pebley and Sastry, 2003). Children living in deprived neighbourhoods are more likely to experience adverse health outcomes when compared with children living in more affluent neighbourhoods (Edwards, 2005). In terms of the built environment, these inequalities may reflect differences in destination access and quality. However, evidence exploring the distribution and quality of resources by neighbourhood disadvantage reveals mixed associations (Timperio *et al.*, 2007; Moore *et al.*, 2008). For example, positive parent- and child-rated neighbourhood quality has been associated with better cognitive, behavioural and emotional outcomes, albeit in girls only (Gagné and Ferrer, 2006), but others found no association with these outcomes in children (Edwards and Bromfield, 2010). Inconsistencies may stem from the different 'neighbourhood quality' measures applied. Neighbourhood quality has been measured in relation to safety and social cohesiveness, which relates to the social environment (Curtis *et al.*, 2004; Bush *et al.*, 2010). Others have considered the quality of specific destinations and services (Edwards and Bromfield, 2010), which relates more to the built environment. Measures of 'quality' can be subjective and complex; however, it appears that both social and physical aspects of the neighbourhood are important and should be considered (Mitchell *et al.*, 2007). Nonetheless, access to high quality child care, developmental programs and education services (e.g. schools) are likely to provide important developmental benefits for children (Pebley and Sastry, 2003; Hertzman, 2004). Friends' houses and recreation areas are other common destinations that children visit (Mackett *et al.*, 2007; Carroll *et al.*, 2015). Green spaces (often parks) have received the most attention in urban health geography, public health and child development literature because it provides a venue for people to be physically active (e.g. children's play, sport participation; Roemmich *et al.*, 2006, Kaczynski *et al.*, 2008), socialise, interact and gather (Maas *et al.*, 2009).

Green spaces and nature

Green spaces and contact with nature aids healthy child development (Pellegrini and Smith, 1998; Wood and Martin, 2010). Studies suggest that natural play environments appear better for children's cognitive and physical development than 'man-made' play areas (see Chapter 12). Children who play in natural areas engage in more physically demanding play compared with traditional playgrounds and demonstrate enhanced gross-motor skills (e.g. climbing) (Fjortoft, 2001), increased attention spans and have fewer sick days at day-care centres (Bagot, 2005). Providing children with a degree of managed risk is important for their development and it teaches children to problem-solve, assess and manage challenging situations, and build resilience, adaptability and self-confidence – qualities that are important for their development and emerging independence (Jones, 2007).

The location and accessibility (via transport and proximity) of green spaces may be important determinants of children and families' decisions to use these destinations in the first instance. Children and adults are more likely to travel to, and use destinations, if they are within walking distance from home (Transportation Research Board, 2005; Page *et al.*, 2010). For younger children, their mobility is primarily dependent on others (e.g. siblings, parents and grandparents) to escort them to places. As children age, their increasing autonomy generally leads to spending more time outside the home without supervision (Leventhal *et al.*, 2009). Local access to, and proximity of, destinations is therefore particularly important because children often travel by foot, scooter, bike or public transport and they are more likely to engage in these modes if locations are close and along a *safe* commute route (Carver *et al.*, 2008).

Traffic

A critical feature of a 'safe commute route' is low traffic exposure. Traffic safety (both child and parent) is the most frequently cited barrier limiting children from being out and about and interacting with their neighbourhood (Panter *et al.*, 2008). Traffic exposure factors that consistently predict children's neighbourhood interaction include traffic levels, speed, design of crossings, illegal or dangerous parking, poor visibility and poor supervision at pedestrian crossings (Kearns *et al.*, 2005; Zwerts *et al.*, 2010).

Residents living on streets with higher traffic volumes may interact less with neighbours compared to those residing on lower trafficked streets (Evans, 2006).

Close proximity to traffic is positively correlated with smaller social networks for children and diminished social skills (Evans, 2006). These reduced interactions likely stem from parental restrictions on children's outdoor play and mobility because of injury fears (Hochschild, 2012). Indeed, child pedestrian and cyclist incidents are a leading cause of death in school-aged children and a major cause of long-term disability (Morris *et al.*, 2001).

For younger children, cul-de-sac networks may facilitate safe play because of lower levels of pedestrian and vehicular traffic (Hochschild, 2012). Moreover, children below the age of 10 years may *not* have the cognitive (e.g. attention focus and interpretation of traffic signs) and perceptual (e.g. judging speed and peripheral vision) abilities to negotiate complex traffic situations (Cross *et al.*, 2000). For older children, interaction and movement through their neighbourhood (particularly when independent) may enhance their development through dealing with risk, learning about their environment and improving spatial and way-finding abilities (e.g. distance estimation, locating north, identifying landmarks) (Rissotto and Tonucci, 2002). Building confidence in a child's ability to navigate their neighbourhood safely through skill-building, and awareness through doing, is therefore important for their spatial and cognitive development.

High-rise density living

As cities rapidly urbanise and shift to higher density development, it is important to consider its potential impact on children's health and development. Although the evidence base is inconclusive and shows mixed associations, some evidence suggests a negative relationship between higher density living and children's developmental outcomes, such as increased behavioural problems (Ineichen and Hooper, 1974), social withdrawal (Evans and Ferguson, 2011) and poorer academic performance (Evans, 2006). The potential mechanisms between residential density and child development are complex. They are likely to be moderated by socio-economic disadvantage and mediated by the negative influence residential density can have on parent mental health, neighbourhood satisfaction and perceptions of the environment (Giles-Corti *et al.*, 2012). These factors appear to be interrelated.

As an example, high-rise housing – greater than three storeys (Newman, 1972) or four storeys (Carroll *et al.*, 2011) – has been positively associated with psychological distress in adults, particularly in low-income mothers of young children (Evans and Ferguson, 2011). Giles-Corti *et al.* (2012) concluded that this may be partly supported by findings demonstrating that families with children younger than five years living in low-rise (rather than high-rise) housing had higher levels of neighbourhood satisfaction (Becker, 1976). This finding is important given that housing and neighbourhood satisfaction is associated with adult mental health, and it is well-recognised that poor parental mental health negatively impacts on a child's development and mental health (Downey and Coyne, 1990). Regardless of socio-economic status and housing quality, parents in families living in higher density neighbourhoods may restrict their children's outdoor play, given that higher density areas tend to attract more traffic and strangers to the vicinity (Villanueva *et al.*,

2013), potentially exacerbating safety concerns and neighbourhood dissatisfaction. Evans and Ferguson (2011) identified that in high-rise developments mothers of young children felt 'trapped' inside and expressed difficulties in monitoring children's outdoor play because of crime and safety concerns. These studies indicate that neighbourhood safety concerns are related to residential density, which in turn affects children's opportunities to interact with others in their neighbourhood to the detriment of their development (Maggi *et al.*, 2010). These findings have implications for designing neighbourhoods with safety in mind, particularly in high-rise residential areas. Indeed, high-rise density can work well if the needs of children and families are considered (City of Vancouver Council, 1992; Carroll *et al.*, 2011). This includes co-locating high-rise development with essential facilities, services and appropriate open space (Whitzman and Mizrachi, 2012) such as parks. Indeed, Min and Lee (2006) interviewed children living in high-rise density housing and found that children find their most psychologically-valued settings in neighbourhood outdoor spaces such as parks.

Significance and challenges for research

Although this area is still evolving, evidence suggests that a number of built environment attributes are related to healthy child development. There is an urgent need to address urban design for healthy child development, given that developmental trajectories are set early and that interventions are most effective in early life. The capacity to progress this interdisciplinary research and develop appropriate interventions that enhance child development has been limited by the lack of conceptual frameworks, spatial data and evidence to support the potential importance of neighbourhood design for child development. However, pursuing this research area is now increasingly possible.

Linking built environment measures with child development outcomes and testing the complexity and strength of these relationships requires multidisciplinary work across various data sources and disciplines (urban planning and policy, urban design, landscape architecture, urban health geography, child development and public health) to determine potential causal mechanisms for these relationships. Integration of subjective (e.g. child and parent perceptions) *and* objective built environment measures is needed to provide a more complete assessment of complex neighbourhood built environments and further refine the child development causal pathways (see Figure 3.1). This approach may offer superior models (Lin and Moudon, 2010) to inform whether interventions need to target the environment, perceptions or both.

Understanding how built environment exposures impact child development is needed to inform the development of neighbourhood indicators relevant to child development. For instance, objective spatial indicators have been developed to test and monitor different environments for the purposes of exploring adult's and children's walking behaviours and neighbourhood destination access (Pearce *et al.*, 2006). As neighbourhood disadvantage has been noted as a predictor of child development (Sampson *et al.*, 2002), exploring these indicators across diverse socio-economic areas and populations – and for different child age groups – will allow comparisons between communities with different physical and social structures. Such work will help identify attributes of places that optimise children's development across the socio-economic spectrum and contribute to reducing health inequalities. Moreover, it will enable communities to better understand their own assets and challenges and also provide guidance to determine the best approaches to optimise children's developmental outcomes. With the current policy environments and global agendas recognising the need to create 'child-friendly' 'liveable' cities and interventions for early childhood development, further research into this area is necessary and timely.

Although there appears to be benefit for neighbourhood interventions to enhance child development outcomes, the potential for 'unintended consequences' should be carefully considered. For example, more walkable neighbourhoods are usually characterised by well-connected streets and a mix of destinations present (Frank *et al.*, 2010). Well-connected streets provide shorter distances to destinations and promote walking; however, they tend to generate more vehicular traffic and increase the road crossing frequency, thereby exposing child pedestrians to greater traffic volume and injury risk potential. Nevertheless, it is also possible that more traffic and pedestrians along popular routes increase children's (or parents') perceived and actual surveillance, which may contribute to feelings of safety (Giles-Corti *et al.*, 2009). Considering a socio-ecological approach to promoting healthy child outcomes is important. With the urban design representing only one factor among many (e.g. child views about their neighbourhood and also social norms) (Mitchell *et al.*, 2007), further interdisciplinary work to unpack these complex relationships is critical and will provide practical guidance for ensuring that interventions work as intended.

Conclusion

A good start to life is critical and lays the foundations for children's full potential as healthy and productive adults. Given that more than 50 per cent of the world's population lives in urban environments, the agenda on building 'liveable' and 'child-friendly' cities is ever more important. Exploring relationships between the neighbourhood built environment and child development will fit another 'piece of the puzzle' and expand the otherwise limited evidence base on why neighbourhoods are important for child development. Such research is needed to inform policy and practice on how to build neighbourhoods that cater for children and families, and develop more effective interventions that encompass the multiple contexts in which children grow and develop.

References

Armstrong, N. 1993. Independent mobility and children's physical development. In: M. Hillman, ed. *Children, transport and the quality of life*. London: Policy Studies Institute, pp. 35–43.

Australian Government Department of Infrastructure and Transport. 2013. *State of Australian cities 2013*. Canberra: Australian Government Department of Infrastructure and Transport.

Badland, H. and Oliver, M. 2012. Child independent mobility: making the case, and understanding how the physical and social environments impact on the behaviour. In: E. Turunen, and A. Koskinen, eds. *Urbanization and the global environment*. New York: NOVA Publishers, pp. 51–79.

Bagot, K. 2005. The importance of green play spaces for children – aesthetic, athletic and academic. *Journal of the Victorian Association for Environmental Education*, 28, pp. 11–15.

Becker, F. 1976. Children's play in multifamily housing. *Environment & Behavior*, 8, pp. 545–74.

Brockman, R., Fox, K. and Jago, R. 2011. What is the meaning and nature of active play for today's children in the uk? *International Journal of Behavioral Nutrition and Physical Activity*, 8, pp. 15–22.

Bronfenbrenner, U. 1979. *The ecology of human development: experiments by nature and design*. Cambridge, MA: Harvard University Press.

Bush, N.R., Lengua, L.J. and Colder, C.R. 2010. Temperament as a moderator of the relation between neighborhood and children's adjustment. *Journal of Applied Developmental Psychology*, 31, pp. 351–61.

Carroll, P., Witten, K. and Kearns, R.A. 2011. Housing intensification in Auckland, New Zealand: implications for children and families. *Housing Studies*, 26, pp. 353–67.

Carroll, P., Witten, K., Kearns, R. and Donovan, P. 2015. Kids in the city: children's use and experiences of urban neighbourhoods in Auckland, New Zealand. *Journal of Urban Design*, 20, pp. 417–36.

Carver, A., Timperio, A. and Crawford, D. 2008. Playing it safe: the influence of neighbourhood safety on children's physical activity – a review. *Health & Place*, 14, pp. 217–27.

Carver, A., Watson, B., Shaw, B. and Hillman, M. 2013. A comparison study of children's independent mobility in England and Australia. *Children's Geographies*, 11, pp. 461–75.

Castonguay, G. and Jutras, S. 2009. Children's appreciation of outdoor places in a poor neighborhood. *Journal of Environmental Psychology*, 29, pp. 101–9.

City of Vancouver Council. 1992. *High-density housing for families with children guidelines*. Vancouver, BC: City of Vancouver Council.

Commission on Social Determinants of Health, 2008. *Closing the gap in a generation: health equity through action on the social determinants of health*. Commission on Social Determinants of Health Final Report. Geneva: World Health Organization.

Cross, D., Stevenson, M., Hall, M., Burns, S., Laughlin, D., Officer, J. and Howat, P. 2000. Child pedestrian injury prevention project: student results. *Preventive Medicine*, 30, pp. 179–87.

Curtis, L.J., Dooley, M.D. and Phipps, S.A. 2004. Child well-being and neighbourhood quality: evidence from the Canadian national longitudinal survey of children and youth. *Social Science & Medicine*, 58, pp. 1917–27.

Downey, G. and Coyne, J.C. 1990. Children of depressed parents: an integrative review. *Psychological Bulletin*, 108, pp. 50.

Edwards, B. 2005. Does it take a village? An investigation of neighbourhood effects on Australian children's development. *Family Matters*, 72, pp. 36–43.

Edwards, B. and Bromfield, L. 2010. Neighbourhood influences on young children's emotional and behavioural problems. *Family Matters*, 84, pp. 7–19.

Evans, G. 2006. Child development and the physical environment. *Annual Review of Psychology*, 57, pp. 423–51.

Evans, G. and Ferguson, K. 2011. Built environment and mental health. In: Jerome O. Nriagu, ed. *Encyclopedia of environmental health*. Burlington, VT: Elsevier, pp. 446–9.

Farrar, E., Goldfeld, S. and Moore, T. 2007. *School readiness. Report for the Australian Research Alliance for Children and Youth*. Melbourne: Children's Research Institute.

Fjortoft, I. 2001. The natural environment as a playground for children: the impact of outdoor play activities in pre-primary school children. *Early Childhood Education Journal*, 29, pp. 111–17.

Frank, L., Sallis, J., Saelens, B., Leary, L., Cain, K., Conway, T. and Hess, P. 2010. The development of a walkability index: application to the neighborhood quality of life study. *British Journal of Sports Medicine*, 44, pp. 924–33.

Freeman, C. and Tranter, P. 2011. *Children and their Urban Environment*. London: Earthscan.

Fyhri, A., Hjorthol, R., Mackett, R.L., Fotel, T.N. and Kyttä, M. 2011. Children's active travel and independent mobility in four countries: development, social contributing trends and measures. *Transport Policy*, 18, pp. 703–10.

Gagné, L.G. and Ferrer, A. 2006. Housing, neighbourhoods and development outcomes of children in Canada. *Canadian Public Policy*, 32, pp. 275–300.

Giles-Corti, B., Kelty, S., Zubrick, S. and Villanueva, K. 2009. Encouraging walking for transport and physical activity in children and adolescents how important is the built environment? *Sports Medicine*, 39, pp. 995–1009.

Giles-Corti, B., Ryan, K. and Foster, S. 2012. *Increasing density in Australia: maximising the health benefits and minimising the harm*. Melbourne: National Heart Foundation of Australia.

Gleeson, B. and Sipe, N. 2006. *Creating child friendly cities: reinstating kids in the city*. London: Routledge.

Herrington, S. and Studtmann, K. 1998. Landscape interventions: new directions for the design of children's outdoor play environments. *Landscape and Urban Planning*, 42, pp. 191–205.

Hertzman, C. 2004. *Making early childhood development a priority: lessons from Vancouver*. Vancouver, BC: Canadian Centre for Policy Alternatives.

Hertzman, C. 2010. *Framework for the social determinants of early child development*. In: Center of Excellence for Early Childhood Development, ed. *Encyclopedia on Early Childhood Development*. Vancouver, Canada: University of British Columbia. Available at www.child-encyclopedia.com/sites/default/files/textes-experts/en/669/framework-for-the-social-determinants-of-early-child-development.pdf (accessed 27 February 2017).

Hochschild, T. 2012. Cul-de-sac kids. *Childhood*, 20(2), pp. 1–15.

Ineichen, B. and Hooper, D. 1974. Wives' mental health and children's behaviour problems in contrasting residential areas. *Social Science & Medicine*, 8, pp. 369–74.

Jacobs, J. 1961. *The death and life of great American cities*. New York: Random House.

Jencks, C. and Mayer, S. 1990. The social consequences of growing up in a poor neighborhood. In: L.E. Lynn Jr and M.G.H. Mcgeary, eds. *Inner city poverty in the United States*. Washington, DC: National Academy Press, pp. 111–86.

Jones, D. 2007. *Cotton wool kids: releasing the potential for children to take risks and innovate*. Coventry: HTI.

Kaczynski, A.T., Potwarka, L.R. and Saelens, B.E. 2008. Association of park size, distance, and features with physical activity in neighborhood parks. *American Journal of Public Health*, 98, pp. 1451.

Kearns, R., Collins, D. and Bean, C. 2005. Children's freedoms and promoting the 'active city' in auckland neighbourhoods. Proceedings of 2nd State of Australian Cities Conference, Brisbane, pp. 1–15.

Kerr, J., Rosenberg, D., Sallis, J., Saelens, B., Frank, L. and Conway, T. 2006. Active commuting to school: associations with environment and parental concerns. *Medicine and Science in Sports and Exercise*, 38, pp. 787–94.

Leventhal, T. and Brooks-Gunn, J. 2000. The neighborhoods they live in: the effects of neighborhood residence on child and adolescent outcomes. *Psychological Bulletin*, 126, pp. 309–37.

Leventhal, T., Dupéré, V. and Brooks-Gunn, J. 2009. Neighborhood influences on adolescent development. In Lerner, R.M. and Steinberg, L. eds. *Handbook of adolescent psychology*. Hoboken, NJ: John Wiley & Sons, Inc., pp. 2:III:12.

Leyden, K. 2003. Social capital and the built environment: the importance of walkable neighborhoods. *American Journal of Public Health*, 93, pp. 1546–51.

Lin, L. and Moudon, A. 2010. Objective versus subjective measures of the built environment, which are most effective in capturing associations with walking? *Health & Place*, 16, pp. 339–48.

Maas, J., Van Dillen, S.M., Verheij, R.A. and Groenewegen, P.P. 2009. Social contacts as a possible mechanism behind the relation between green space and health. *Health & Place*, 15, pp. 586–95.

Mackett, R., Brown, B., Gong, Y., Kitazawa, K. and Paskins, J. 2007. Children's independent movement in the local environment. *Built Environment*, 33, pp. 454–68.

Maggi, S., Irwin, L., Siddiqi, A. and Hertzman, C. 2010. The social determinants of early child development: an overview. *Journal of Paediatrics and Child Health*, 46, pp. 627–35.

Min, B. and Lee, J. 2006. Children's neighborhood place as a psychological and behavioral domain. *Journal of Environmental Psychology*, 26, pp. 51–71.

Mitchell, H., Kearns, R. and Collins, D. 2007. Nuances of neighbourhood: children's perceptions of the space between home and school in Auckland, New Zealand. *Geoforum*, 38, pp. 614–627.

Moore, L., Diez Roux, A., Evenson, K., Mcginn, A. and Brines, S. 2008. Availability of recreational resources in minority and low socioeconomic status areas. *American Journal of Preventive Medicine*, 34, pp. 16–22.

Morris, J., Wang, F. and Lilja, L. 2001. School children's travel patterns – a look back and a way forward. *Transport Engineering in Australia*, 7, pp. 15–25.

Newman, O. 1972. *Defensible space*. London: Architectural Press.

Nicholson, J.M., Lucas, N., Berthelsen, D. and Wake, M. 2012. Socioeconomic inequality profiles in physical and developmental health from 0–7 years: Australian national study. *Journal of Epidemiology & Community Health*, 66, pp. 81–7.

Page, A., Cooper, A., Griew, P. and Jago, R. 2010. Independent mobility, perceptions of the built environment and children's participation in play, active travel and structured exercise and sport: the Peach Project. *International Journal of Behavioral Nutrition and Physical Activity*, 7, pp. 17.

Panter, J., Jones, A. and Van Sluijs, E. 2008. Environmental determinants of active travel in youth: a review and framework for future research. *International Journal of Behavioral Nutrition and Physical Activity*, 5, pp. 34–48.

Pearce, J., Witten, K. and Bartie, P. 2006. Neighbourhoods and health: a GIS approach to measuring community resource accessibility. *Journal of Epidemiology and Community Health*, 60, pp. 389–95.

Pebley, A. and Sastry, N. 2003. *Concentrated poverty vs. Concentrated affluence: effects on neighborhood social environments and children's outcomes.* Paper presented at the Annual Meeting of the Population Association of America, Minneapolis, MN, 1 May 2003.

Pellegrini, A.D. and Smith, P.K. 1998. Physical activity play: the nature and function of a neglected aspect of play. *Child Development*, 69, pp. 577–98.

Prezza, M. and Pacilli, M. 2007. Current fear of crime, sense of community, and loneliness in italian adolescents: the role of autonomous mobility and play during childhood. *Journal of Community Psychology*, 35, pp. 151–70.

Rissotto, A. and Tonucci, F. 2002. Freedom of movement and environmental knowledge in elementary school children. *Journal of Environmental Psychology*, 22, pp. 65–77.

Roemmich, J.N., Epstein, L.H., Raja, S., Yin, L., Robinson, J. and Winiewicz, D. 2006. Association of access to parks and recreational facilities with the physical activity of young children. *Preventive Medicine*, 43, pp. 437–41.

Sallis, J., Prochaska, J. and Taylor, W. 2000. A review of correlates of physical activity of children and adolescents. *Medicine and Science in Sports Exercise*, 32, pp. 963–75.

Sallis, J.F., Floyd, M.F., Rodríguez, D.A. and Saelens, B.E. 2012. Role of built environments in physical activity, obesity, and cardiovascular disease. *Circulation*, 125, pp. 729–37.

Sampson, R., Morenoff, J. and Gannon-Rowley, T. 2002. Assessing 'neighbourhood effects': social processes and new directions in research. *Annual Review Sociology*, 28, pp. 443–78.

Timperio, A., Ball, K., Salmon, J., Roberts, R. and Crawford, D. 2007. Is availability of public open space equitable across areas? *Health & Place*, 13, pp. 335–40.

Transportation Research Board. 2005. *Does the built environment influence physical activity? Examining the evidence.* Washinton, DC: Transportation Research Board.

Tranter, P. and Whitelegg, J. 1994. Children's travel behaviours in Canberra: car-dependent lifestyles in a low-density city. *Journal of Transport Geography*, 2, pp. 265–73.

United Nations. 2010. *World urbanisation prospects: The 2009 revision.* New York: United Nations Department of Economic and Social Affairs: Population Division.

UNICEF. 2013. *ECD in social policies.* Available at www.unicef.org/earlychildhood/index_69848.html (accessed 1 September 2013).

Villanueva, K., Giles-Corti, B., Bulsara, M., Timperio, A., Mccormack, G., Beesley, B., Trapp, G. and Middleton, N. 2013. Where do children travel to and what local opportunities are available? The relationship between neighborhood destinations and children's independent mobility. *Environment and Behavior*, 45, pp. 679–705.

Whitzman, C. and Mizrachi, D. 2012. Creating child-friendly high-rise environments: beyond wastelands and glasshouses. *Urban Policy and Research*, pp. 1–17.

Wood, L. and Giles-Corti, B. 2008. Is there a place for social capital in the psychology of health and place? *Journal of Environmental Psychology*, 28, pp. 154–63.

Wood, L. and Martin, K. 2010. *What makes a good play area for children?* Perth, Australia: Centre for the Built Environment and Health, University of Western Australia.

Wood, L., Frank, L. and Giles-Corti, B. 2010. Sense of community and its relationship with walking and neighborhood design. *Social Science & Medicine*, 70, pp. 1381–90.

Wood, L., Giles-Corti, B., Zubrick, S.R. and Bulsara, M.K. 2013. 'Through the kids. . . we connected with our community' children as catalysts of social capital. *Environment and Behavior*, 45, pp. 344–68.

Zubrick, S., Wood, L., Villanueva, K., Wood, G., Giles-Corti, B. and Christian, H. 2010. *Nothing but fear itself: parental fear as a determinant of child physical activity and independent mobility*. Melbourne: Victorian Health Promotion Foundation.

Zwerts, E., Allaert, G., Janssens, D., Wets, G. and Witlox, F. 2010. How children view their travel behaviour: a case study from Flanders (Belgium). *Journal of Transport Geography*, 18, pp. 702–10.

4 Changing geographies of children's air pollution exposure

*Shanon Lim, Jennifer A. Salmond
and Kim N. Dirks*

One of the most significant public health challenges of the twenty-first century is managing and mitigating the impact of poor air quality on human mortality, morbidity, comfort and quality of life. Air pollution in particular affects the air–tissue interface of the lung resulting in wheezing, coughing, pneumonia, acute bronchitis and asthma exacerbations of those exposed (Kulkarni and Grigg, 2008). Due to increasing vehicle ownership rates and the regulation of industrial and domestic heating emissions, the dominant source of air pollution in many developed cities is transport related (Kaur *et al.*, 2007). Health studies have shown increased morbidity in individuals living close to major roads and transportation networks (Brauer *et al.*, 2006; McConnell *et al.*, 2006; Morawska *et al.*, 2009). More recently, studies have also shown that the highest pollution exposures occur within the transport (often denoted 'commuter') micro-environment (Buonanno *et al.*, 2013).

Children often have little control over their exposure to air pollution and are disproportionately affected by poor air quality (World Health Organization, 2015). Exposure to air pollutants in early life, including during the pre-natal phase, can have lifelong negative consequences for individual health (Salvi, 2007; Kulkarni and Grigg, 2008; Tang *et al.*, 2014). Children are particularly vulnerable because their lungs are still developing and they lack the defence functions found in adults (Salvi, 2007; Ashmore and Dimitroulopoulou, 2009). Children also have increased breathing rates and inhale a higher volume of air relative to their body weight compared to adults (Ginsberg *et al.*, 2005; Buonanno *et al.*, 2012).

Transport-related emissions are of particular concern for children who spend large proportions of their days at schools located close to main highways. Irrespective of the location, children are encouraged to spend time exercising outdoors, potentially resulting in increased exposure (Buonanno *et al.*, 2012; Lin, 2013; Clelland *et al.*, 2015). However, despite the known health risks, our ability to quantify and understand the health effects of children's exposure to air pollutants remains limited.

This chapter focuses on the changing geographies of children's exposure to air pollution in developed world cities where air pollution is predominantly derived from transport emissions. We use the term 'air pollution' to refer to the presence of an airborne substance that exceeds a level such that it is harmful to human health. Our focus is on common transport-related pollutants, including particulate matter (ultrafine particulates (UFP) and particulate matter less than 10 and 2.5 microns in diameter), nitrogen dioxide, carbon monoxide and volatile organic compounds such as polycyclic aromatic hydrocarbons and sulphur dioxide.

The aim of this chapter is to demonstrate how the exposure of children to air pollution changes with age and also in time and space. Traditional air pollution studies have used fixed outdoor monitors that examine the relationship between health and place over large temporal and spatial scales. This work has formed an important basis for understanding the impacts of air pollution on children. However, such an approach has a number of limitations due to children's mobility. More recent work on adult exposure to air pollution has focused on local-scale studies where air pollution monitors are attached to participants and personal exposure is tracked through time and space (Fang and Lu, 2012). Although such studies also have limitations, they have shown the importance of quantifying the time spent, activity undertaken and individual behaviours in different microenvironments in determining exposure and dose of air pollution (Bruinen de Bruin *et al.*, 2004). Such studies have also highlighted the importance of both the indoor and commuter microenvironments (Cattaneo *et al.*, 2009; Buonanno *et al.*, 2013). We suggest that although to date this approach to monitoring personal exposures has only limited uptake in the literature regarding children, understanding the ways in which children move through the polluted environment may be fundamental in determining their daily exposure pathways, and ultimately mitigating the impact of poor air quality. We also suggest that there is scope for exploring children's understanding of their own exposure to air pollution and the role of education as a tool for mitigating exposure.

Epidemiological approaches to determining children's health effects from air pollution

Traditional epidemiological research is undertaken by analysing long-term air quality monitoring data (at a daily, monthly or annual scale) measured as close as possible to the place of residence and correlating this information with large health datasets. To assess the varying effects of ambient air pollution exposure, participants are recruited from different geographic exposure areas, controls are introduced for confounding socio-economic status or health factors (underlying disease, diet, genetics, etc.) and each population is analysed for any adverse impact of air quality (Kulkarni and Grigg, 2008).

Epidemiological studies using this approach have found significant correlations between ambient pollution and poor health amongst children (Gauderman *et al.*, 2004; Salvi, 2007; Kulkarni and Grigg, 2008; Tang *et al.*, 2014). In particular, exposure to ambient air pollution has been shown to significantly increase the risk of respiratory diseases in children (Weichenthal *et al.*, 2006; Salvi, 2007). For example, a review of the international literature found that 16 per cent of all respiratory-related hospital admissions amongst new-born children are associated with air pollution (Salvi, 2007). One of the largest studies of its kind monitored 1800 children for eight years in southern California and found reduced lung function when children lived in areas of higher levels of nitrogen dioxide and particulate matter (Gauderman *et al.*, 2004). Similar work in Japan found a small association between the incidence of asthma in children and ambient particulate concentrations less than 10 microns (Shima *et al.*, 2002).

Studies have also determined that air pollution has the potential to affect children whilst they are still in utero. For example, lower birth weight and higher incidences of premature birth have been found amongst mothers living in areas associated with higher levels of nitrogen dioxide in a number of different countries (Kulkarni and Grigg, 2008). Tang *et al.* (2014) found increased birth weight and growth in infants born after a nearby polluting power plant in Tongliang, China, was shut down compared to infants in the same community born while the power plant was in operation. Other studies have found a fivefold increase in cough and fourfold increase in wheeze during the first year of an infant's life when pregnant mothers were exposed to high ambient levels of polycyclic aromatic hydrocarbons during pregnancy (Salvi, 2007).

Early studies investigating the impact of air pollution on physiology were done by measuring the physiology of subjects in response to exposure to large doses of air pollution in a lab-based environment (e.g. carbon monoxide). For what were considered ethical reasons at the time, much of this work was carried out only on young adult men (e.g. Coburn *et al.*, 1965; Peterson and Stewart, 1970). Later studies included women (e.g. Peterson and Stewart, 1975). However, even today, ethical considerations are such that studies rarely focus specifically on children (Weichenthal *et al.*, 2006), and therefore there remain gaps in our understanding in terms of the extent to which what we know about the physiological response to air pollution exposure in adults can be extrapolated to that in children, especially the very young, given their different body size and physiology.

Proximity to traffic and associated health effects

Clearly such traditional approaches to identifying the health impacts of air pollution on children have a number of limitations. One of the key limiting factors to any study that uses air quality data from a single fixed monitor is the assumption that ambient concentrations are homogeneous in space. Studies have shown that concentrations of traffic-related pollutants drop quickly as you move away from major road sources, and intra-urban measurements of air pollution have demonstrated the spatial and temporal heterogeneity in air pollution concentrations within urban areas (Kaur *et al.*, 2005). It is therefore very unlikely that a single measurement can represent air quality typical of the large residential areas or census units used in these studies, especially when dealing with primarily traffic-related pollutant exposure measurements (Asmi *et al.*, 2009; Molle *et al.*, 2013).

Fixed monitors do not take into account the mobility of individuals and neglect the indoor and commuting micro-environment where air pollution concentrations can be significantly different from levels observed outdoors (Buonanno *et al.*, 2013). Therefore, although these studies have provided valuable information on the possible health effects of air pollution, they reveal little in the way of an accurate representation of children's actual exposure (Gauderman *et al.*, 2004; Weichenthal *et al.*, 2006; Morawska *et al.*, 2009; Buonanno *et al.*, 2013).

Studies that have provided higher resolution spatial analysis of pollutant concentrations and health have demonstrated the importance of the proximity to the road sources in determining the health outcomes of children. These studies have shown that the location of a child's home, school or early childhood centre (ECC) in relation to their proximity to busy roads is crucial for predicting children's air pollution exposure and the associated adverse health effects (Brauer *et al.*, 2006; McConnell *et al.*, 2006; Brugge *et al.*, 2007; Morawska *et al.*, 2009; Clelland *et al.*, 2015). A study by Clelland *et al.* (2015) found that ECCs located closer to high-traffic roads have higher ambient levels of nitrogen dioxide (see Figure 4.1). The study also found that outdoor areas at ECCs were typically located on the side of the ECC closer to the main road instead of behind the building. This resulted in higher levels of nitrogen dioxide in outdoor play areas where children are most active.

These results suggest that understanding where children are located in space at various times of the day and on different days of the week is important in determining their exposure, and reliable estimates of personal exposure need to utilise personal mobile monitoring techniques (Alm *et al.*, 1994; Ashmore and Dimitroulopoulou, 2009; Mejía *et al.*, 2011).

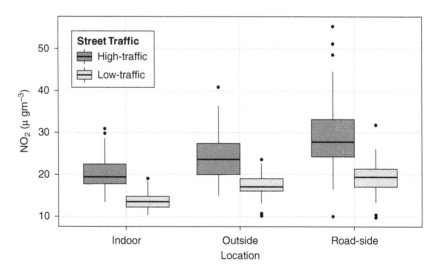

Figure 4.1 Comparison of nitrogen dioxide (NO_2) concentrations between indoor, outside and road-side locations of high-traffic and low-traffic early childhood centres.

Source: adapted from Clelland *et al.*, 2015.

Children's pollution exposure over time and space

Air pollution exposure studies involving mobile monitors typically consider three air pollution variables: exposure (defined here as the average amount of air pollution that a child is exposed to); dose (defined as the product of exposure and time, typically represented as a percentage of day); and inhaled dose (an estimate of the amount of pollution a child inhales, a product of exposure, time and their inhalation rate). The latter is most important in children's exposure studies because the inhalation rate is significantly higher than that of adults relative to body weight, especially because children tend to engage in higher levels of physical activity and have higher resting breathing rates.

Mobile monitoring studies give greater detail of exposures in time and space compared to those of fixed monitors. However, to date, there have only been a limited number of daily exposure studies of children that can be used to identify the most influential microenvironments and the times when children's air pollution exposure is at its highest. From these studies, three main environments have been identified in which significant exposure to air pollution occurs for children: the school environment, at home and during the daily commute. Therefore, mobile monitoring research in recent years has focused on these spaces.

Pollution exposure during commuting

As with adults, the commute (in this case to and from school or ECC) often represents the highest exposure during a child's day (Alm *et al.*, 1994; Buonanno *et al.*, 2012; Lin, 2013). Commuting exposures for children have been found to be significantly higher during the morning commute compared to the afternoon commute, partly because of the reduced dispersion and also because of the coincidence of the morning school and work commutes (Lin, 2013). Figure 4.2 also shows that children experience significantly higher average exposures to UFP during their commute times compared to during the time spent at school.

Different modes of commuting are known to be associated with different exposures and doses of air pollution in adults (Dirks *et al.*, 2012). Current policy and culture encourage active modes of commuting, such as walking and cycling, because of the physical and mental health benefits. In children, walking to school is also considered beneficial from a developmental point of view (Collins and Kearns, 2005; Kullman, 2010). Walking has been found to result in the lowest pollutant exposure compared to motorised commute modes in adults (Kaur *et al.*, 2007; McNabola *et al.*, 2009; Grange *et al.*, 2014). However, these studies tend not to take into account breathing rate or the increased travel time, which could significantly increase inhaled doses compared to sedentary commute modes. Studies of adults have found that inhaled doses from walking can be up to 10 times higher compared to those of adult car commuters in a variety of cities around the world (Mudu *et al.*, 2006) including Cassino, Italy (Buonanno *et al.*, 2012) and Auckland, New Zealand (Dirks *et al.*, 2012). This value is likely to be significantly higher for children but has yet to be studied.

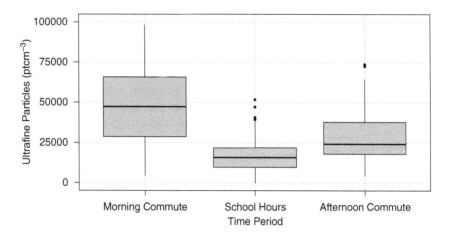

Figure 4.2 Ultrafine particles exposure over the morning commute, school hours and afternoon commute for children at a school near high-traffic roads in Auckland, New Zealand.

Source: adapted from Lin, 2013.

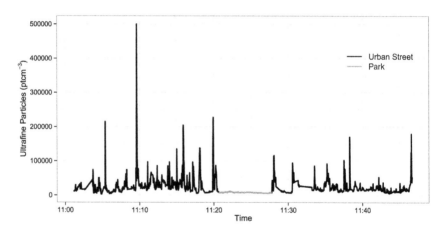

Figure 4.3 Time series of ultrafine particle exposure when moving through urban and green spaces in Auckland, New Zealand.

Lin (2013) studied the walk to and from school in urban areas and the effect of route choice. The study found lower pollutant exposure on low trafficked streets, and particularly high exposures when crossing under motorway bridges in Auckland, New Zealand. Goodey (2011) found a significant drop in UFP exposure when walking through green spaces compared to urban streets (Figure 4.3).

Travelling times and the frequency of children commuting by cars has risen in recent years, therefore increasing their pollution exposure. There are a number of factors which can influence pollution exposure while travelling in a vehicle including fuel type, air conditioning settings, vehicle type, and congestion (Ashmore and Dimitroulopoulou, 2009). Children are also frequently exposed to vehicular exhaust through school bus transportation. A typical child may ride a school bus 180 days in a year. As such, there has been a number of studies carried out on school bus transportation, particularly in the US (Zhang and Zhu, 2010). In recent years, there have been significant improvements in bus technology and management such as retrofits, cleaner fuel and reduced idling times; this has resulted in decreased pollution exposure on buses (Gao and Klein, 2011; Zhang and Zhu, 2011). Despite these improvements, bus transportation still contributes significantly to a school child's daily dose of pollution (Zhang and Zhu, 2010), and a tendency towards using older fleet vehicles on school bus routes exacerbates this effect.

Influence of indoor air pollution

Pollution exposure during the commute has long been the sole focus of mobile air pollution studies, due to the fact that the majority of pollution recorded by fixed monitors is transport related. However, the dominance of exposure in the commuter microenvironment in determining daily pollutant exposure and dose amongst children is not universal. It is influenced by the type of pollutant under consideration and the characteristics of the indoor environments. For example, an exposure study of 100 participants in Cassino, Italy identified the highest dose to UFP in the home environment, accounting for 57 per cent of children's daily exposure followed by 18 per cent in the school environment (Buonanno *et al.*, 2012). Similar results were found in Brisbane, Australia where the home environment was found to contribute 55 per cent of the total inhaled dose of UFP, the school environment 35 per cent and, surprisingly, the commute environment only 4.5 per cent (Mazaheri *et al.*, 2014). Overall, the inhaled doses of UFP were found to be up to five times higher in Italy compared to Australia (Buonanno *et al.*, 2012; Mazaheri *et al.*, 2014), indicating national differences in exposure.

   The importance of the indoor environment is increasingly being recognised, especially as on average, 85 per cent of a person's day is spent indoors (Bruinen de Bruin *et al.*, 2004). The majority of time spent by children in a single location (on average over 60 per cent of a school child's day) is within their own home (Mejía *et al.*, 2011). The prevalence of television and gaming devices also limits the amount of time spent on outdoor activities for some cohorts of children (Ashmore and Dimitroulopoulou, 2009). Cooking and eating activities have been found to result in the highest inhaled doses of air pollution in the home environment (Hussein *et al.*, 2006; Mazaheri *et al.*, 2014). Other pollutant sources in this space include biomass burning, smoking, air fresheners, and cleaning products (Alm *et al.*, 1994; Buonanno *et al.*, 2013).

   On average, 25 per cent of a child's day is spent at school, with the majority of this time spent indoors (Laiman *et al.*, 2014). In recent years, there has been an increasing number of exposure studies undertaken in schools (Morawska *et al.*, 2009;

Mejía *et al.*, 2011; Buonanno *et al.*, 2013). In indoor school areas, art, printing, and cleaning products contribute significantly to classroom air pollution (Morawska *et al.*, 2009). Other sources of air pollution in schools include grilling (use of BBQs) outside, which could infiltrate classrooms and have been shown to increase the exposure of children within the classroom in schools in Brisbane, Australia (Guo *et al.*, 2010; Laiman *et al.*, 2014). Studies have shown that classrooms are often poorly ventilated, leaving pollutants trapped inside for long periods of time (Mejía *et al.*, 2011). The lifetime of pollution events in poorly ventilated classrooms could be expected to be two to three hours (Laiman *et al.*, 2014).

The location of schools has also been found to affect children's pollution exposure. Buonanno *et al.* (2012) found that urban school students experienced 25 per cent higher exposure to UFP compared to those attending rural schools, due largely to the home and school environments. Lin (2013) investigated three different school locations in Auckland. The study found that children with the highest UFP exposure were those attending a central city school close to busy roads and motorways, whereas children with the lowest exposure attended a suburban school located over 1 km from a main road.

Although there is significant benefit in undertaking mobile monitoring research, there are also challenges – the equipment required to run field campaigns is expensive and resource-intensive to deploy. Moreover, the significant person-to-person variability in exposures requires large sample sizes to ensure adequate representation. Sampling involving children also presents an additional challenge to ensure that the activities of the child carrying the equipment are not compromised and that the equipment remains safe from damage. As such, only a few studies have specifically focused on children and it is not possible to infer exposures accurately from similar studies of adults (Alm *et al.*, 1994; Buonanno *et al.*, 2012; Lin, 2013; Mazaheri *et al.*, 2014). Children's personal exposure pathways to air pollution therefore remain grossly understudied, particularly in light of their vulnerability. Further mobile monitoring studies are needed to improve our understanding of children's air pollution exposure (Ashmore and Dimitroulopoulou, 2009; Mejía *et al.*, 2011).

Children's awareness and control over exposure to air pollution

There have been very few studies of children's understanding, awareness and response to air pollution. Children often do not have control over what air pollution they are exposed to and may lack a voice in controlling key decisions as to how they move through space (Holloway, 2014). Even when they do have choices, such as where to play on the playing field or which route to take to school, many children do not have a basic understanding of the implications for their exposure or health. Children may experience some vehicle related pollution negatively through noxious smells or coloured air, but other pollutants such as carbon monoxide have little smell or colour, which makes it difficult for them to be identified.

In one of a limited number of studies of children in British schools, 11–16-year-olds showed an acute awareness of the dangers of being struck by vehicles (Batterham *et al.*, 1996). However, Batterham *et al.* (1996) note that their understanding of the health and environmental impact was poorly conceptualised and often linked to misunderstandings regarding global-scale environmental impacts such as the destruction of the ozone layer. Older children were better at identifying a link between vehicle-related pollution and respiratory health (Batterham *et al.*, 1996). Although even 10–11-year-olds were able to suggest general ways to reduce exhaust emissions, few of the children studied were able to relate their own actions to either reducing local pollution emissions or their own exposure to air pollution (Leeson *et al.*, 1997). Overall, children's (particularly primary school age and below) ideas about local air pollution remain largely unexplored using qualitative or quantitative techniques. Given their 'intrinsic interest in the environment' (Batterham *et al.*, 1996: 347), improved education may provide children with more effective voices and more control over their own exposures, helping to mitigate their exposure to air pollution.

Conclusion

The geography of children's exposure to air pollution is constantly changing. Children's pollution exposure is variable in both time and space, over continents, cities and microenvironments, minute by minute, throughout the day, as well as seasonally. The effects of exposure from air pollution differ depending on age, activity and time spent in different environments. However, children are disproportionately affected compared to adults due to their physical vulnerability and the cumulative health effects of air pollution over time as children move to their adult years. Children often do not have control over the air pollution they are exposed to and may lack a voice in controlling key decisions that determine their exposure. All of these factors mean that air pollution exposure may have a significant effect on children's health and quality of life, both in the present and in the future.

Over time, our understanding of air pollution exposure has changed with a move away from outdoor fixed monitors for predicting a population's exposure to studies involving real-time mobile monitors for recording the personal exposure of individuals in key microenvironments. However, our understanding of the consequences of children's mobilities for their exposure to air pollution in different microenvironments and the resulting health effects of pollution remain limited. This is due to shortcomings with both fixed and mobile monitor studies. Challenges remain both with reducing children's pollution exposure and obtaining accurate exposure datasets in order to improve our understanding of children's exposure.

There are a number of approaches that could be taken to reduce exposure, including adult and child education, leading to changes in behaviour, as well as legislative changes influencing behaviour. As children are now spending the majority of their day indoors, in schools and in homes, measures need to be put in place to educate adults about reducing children's pollution exposure indoors

(McIntosh *et al.*, 2015). Restrictions on activities such as cleaning facilities outside school hours, having printers and art classes in separate areas away from main classrooms, and improving the ventilation systems of both school and home indoor areas will all help to reduce pollutant exposures. Additionally, the use of modern heating appliances such as heat pumps instead of biomass burning or gas heaters will also improve indoor air quality. The location of microenvironments such as schools, early childhood centres and houses needs to be evaluated with respect to distance from heavily trafficked routes. Co-location of adult with child education facilities may be one of the most effective options for improving exposure awareness and allowing individuals to mitigate their own exposure. Additional law enforcement may also be beneficial, for example prohibiting smoking in vehicles that are transporting children (Tymko and Collins, 2014), promoting the use of modern buses on school routes and limiting the use of unflued gas heaters would all be beneficial.

There is also a need to increase these restrictions to outdoor air pollution sources by creating pedestrian-only streets, limiting traffic around schools and providing good access to public transport. Improving vehicle emission technology will help reduce the amount of pollution produced, as long as increases in vehicle numbers do not offset benefits. The promotion and provision of walking school buses (which use low pollution routes) is another example of measures that can be taken to provide the infrastructure required to reduce personal exposure for children, especially because this reduces congestion around school zones (Collins and Kearns, 2005). In older children, education in the area of air pollution exposure linked to travel to school (a better understanding of the consequences of travel mode and route choice) may be highly beneficial in reducing exposure. Identifying ways to encourage active modes of commuting to and from school, particularly if the route moves through green spaces, will also benefit the mental and physical health of our children.

In summary, the geographies of children's exposure to air pollution are highly variable in time and space. Studies have consistently identified the detrimental health effects to children of air pollution. However, an improved understanding of how children are exposed to air pollutants as they move through space at local scales is required to develop some clear guidance on steps that can be taken, at a personal and policy level, to reduce exposure and ensure a healthier future for our children.

References

Alm, S., Reponen, A., Mukala, K., Pasanen, P., Tuomisto, J. and Jantunen, M.J. 1994. Personal exposure of preschool children to carbon monoxide: roles of ambient air quality and gas stoves. *Atmospheric Environment*, 28(22), pp. 3577–80.
Ashmore, M.R. and Dimitroulopoulou, C. 2009. Personal exposure of children to air pollution. *Atmospheric Environment*, 43(1), pp. 128–41.
Asmi, E., Antola, M., Yli-Tuomi, T., Jantunen, M., Aarnio, P., Makela, T., Hillamo, R. and Hameri, K. 2009. Driver and passenger exposure to aerosol particles in buses and trams in Helsinki, Finland. *Science of the Total Environment*, 407(8), pp. 2860–67.

Batterham, D., Stanisstreet, M. and Boyes, E. 1996. Kids, cars and conservation: children's ideas about the environmental impact of motor vehicles. *International Journal of Science Education*, 18(3), pp. 347–54.

Brauer, M., Gehring, U., Brunekreef, B., de Jongste, J., Gerritsen, J., Rovers, M., Wichmann, H.-E., Wijga, A. and Heinrich, J. 2006. Traffic-related air pollution and otitis media. *Environmental Health Perspectives*, 114(9), pp. 1414–18.

Brugge, D., Durant, J.L. and Rioux, C. 2007. Near-highway pollutants in motor vehicle exhaust: a review of epidemiologic evidence of cardiac and pulmonary health risks. *Environmental Health*, 6, p. 23.

Bruinen de Bruin,Y., Carrer, P., Jantunen, M., Hanninen, O., Di Marco, G.S., Kephalopoulos, S., Cavallo, D. and Maroni, M. 2004. Personal carbon monoxide exposure levels: contribution of local sources to exposures and microenvironment concentrations in Milan. *Journal of Exposure Analysis and Environmental Epidemiology*, 14(4), pp. 312–22.

Buonanno, G., Marini, S., Morawska, L. and Fuoco, F.C. 2012. Individual dose and exposure of Italian children to ultrafine particles. *Science of the Total Environment*, 438, pp. 271–7.

Buonanno, G., Marks, G.B. and Morawska, L. 2013. Health effects of daily airborne particle dose in children: direct association between personal dose and respiratory health effects. *Environmental Pollution*, 180, pp. 246–50.

Cattaneo, A., Garramone, G., Taronna, M., Peruzzo, C. and Cavallo, D.M. 2009. Personal exposure to airborne ultrafine particles in the urban area of Milan. *Journal of Physics: Conference Series*, 151(1), 012039.

Clelland., P., Lyne, M., Salmond, J.A., Chelimo, C. and Dirks, K.N. 2015. Nitrogen dioxide exposure at early childhood centres in proximity to high and low traffic roads in Auckland, New Zealand. *Air Quality and Climate Change*, Feb, pp. 28–31.

Coburn, R.F., Forster, R.E. and Kane, P.B. 1965. Considerations of the physiological variables that determine the blood carboxyhemoglobin concentration in man. *Journal of Clinical Investigation*, 44, pp. 1899–910.

Collins, D.C.A. and Kearns, R.A. 2005. Geographies of inequality: child pedestrian injury and walking school buses in Auckland, New Zealand. *Social Science & Medicine*, 60(1), pp. 61–9.

Dirks, K.N., Sharma, P., Salmond, J.A. and Costello, S.B. 2012. Personal exposure to air pollution for various modes of transport in Auckland, New Zealand. *The Open Atmospheric Science Journal*, 6, pp. 84–92.

Fang, T.B. and Lu, Y. 2012. Personal real-time air pollution exposure assessment methods promoted by information technological advances. *Annals of GIS*, 18(4), pp. 279–88.

Gao, H.O. and Klein, R.A. 2011. Environmental equity in funding decisions of the Clean Air School Bus Program: the case of New York State. *Transportation Research Part D: Transport and Environment*, 16(1), pp. 10–14.

Gauderman, W.J., Avol, E., Gilliland, F., Vora, H., Thomas, D., Berhane, K., McConnell, R., Kuenzli, N., Lurmann, F., Rappaport, E., Margolis, H., Bates, D. and Peters, J. 2004. The effect of air pollution on lung development from 10 to 18 years of age. *The New England Journal of Medicine*, 351(11), pp. 1057–67.

Ginsberg, G.L., Foos, B.P. and Firestone, M.P. 2005. Review and analysis of inhalation dosimetry methods for application to children's risk assessment. *Journal of Toxicology and Environmental Health, Part A*, 68(8), pp. 573–615.

Goodey, H. 2011. Ultrafine particle exposure of adult and child pedestrians in urban environments. Masters thesis, University of Auckland.

Grange, S.K., Dirks, K.N., Costello, S.B. and Salmond, J.A. 2014. Cycleways and footpaths: what separation is needed for equivalent air pollution dose between travel modes? *Transportation Research Part D: Transport and Environment*, 32, pp. 111–19.

Guo, H., Morawska, L., He, C., Zhang, Y.L., Ayoko, G. and Cao, M. 2010. Characterization of particle number concentrations and Pm2.5 in a cchool: influence of outdoor air pollution on indoor air. *Environmental Science and Pollution Research*, 17(6), pp. 1268–78.

Holloway, S.L. 2014. Changing children's geographies. *Children's Geographies*, 12(4), pp. 377–92.

Hussein, T., Glytsos, T., Ondráček, J., Dohányosová, P., Ždímal, V., Hämeri, K., Lazaridis, M., Smolík, J. and Kulmala, M. 2006. Particle size characterization and emission rates during indoor activities in a house. *Atmospheric Environment*, 40(23), pp. 4285–307.

Kaur, S., Nieuwenhuijsen, M. and Colvile, R. 2005. Personal exposure of street canyon intersection users to Pm2.5, ultrafine particle counts and carbon monoxide in Central London, UK. *Atmospheric Environment*, 39(20), pp. 3629–41.

Kaur, S., Nieuwenhuijsen, M.J. and Colvile, R.N. 2007. Fine particulate matter and carbon monoxide exposure concentrations in urban street transport microenvironments. *Atmospheric Environment*, 41(23), pp. 4781–810.

Kulkarni, N. and Grigg, J. 2008. Effect of air pollution on children. *Paediatrics and Child Health*, 18(5), pp. 238–43.

Kullman, K. 2010. Transitional geographies: making mobile children. *Social & Cultural Geography*, 11(8), pp. 829–46.

Laiman, R., He, C., Mazaheri, M., Clifford, S., Salimi, F., Crilley, L.R., Megat Mokhtar, M.A. and Morawska, L. 2014. Characteristics of ultrafine particle sources and deposition rates in primary schoo classrooms. *Atmospheric Environment*, 94, pp. 28–35.

Leeson, E., Stanisstreet, M. and Boyes, E. 1997. Primary children's ideas about cars and the environment. *Education 3–13: International Journal of Primary, Elementary and Early Years Education*, 25(2), pp. 25–9.

Lin, H. 2013. Exposure of school children to traffic-related ultrafine particles during travel to school and school hours. Masters thesis, University of Auckland.

Mazaheri, M., Clifford, S., Jayaratne, R., Megat Mokhtar, M.A., Fuoco, F., Buonanno, G. and Morawska, L. 2014. School children's personal exposure to ultrafine particles in the urban environment. *Environmental Science and Technology*, 48(1), pp. 113–20.

McConnell, R., Berhane, K., Yao, L., Jerrett, M., Lurmann, F., Gilliland, F., Künzli, N., Gauderman, J., Avol, E., Thomas, D. and Peters, J. 2006. Traffic, susceptibility, and childhood asthma. *Environmental Health Perspectives*, 114(5), pp. 766–72.

McIntosh, A.H., Collins, D. and Parsons, M. 2015. 'A place for healthy activity': parent and caregiver perspectives on smokefree playgrounds. *Health Place*, 31, pp. 146–53.

McNabola, A., Broderick, B.M. and Gill, L.W. 2009. A principal components analysis of the factors effecting personal exposure to air pollution in urban commuters in Dublin, Ireland. *Journal of Environmental Science and Health. Part A, Toxic/Hazardous Substances and Environmental Engineering*, 44(12), pp. 1219–26.

Mejía, J.F., Choy, S.L., Mengersen, K. and Morawska, L. 2011. Methodology for assessing exposure and impacts of air pollutants in school children: data collection, analysis and health effects – a literature review. *Atmospheric Environment*, 45(4), pp. 813–23.

Molle, R., Mazoué, S., Géhin, É. and Ionescu, A. 2013. Indoor–outdoor relationships of airborne particles and nitrogen dioxide inside Parisian buses. *Atmospheric Environment*, 69, pp. 240–8.

Morawska, L., He, C., Johnson, G., Guo, H., Uhde, E. and Ayoko, G. 2009. Ultrafine particles in indoor air of a school: possible role of secondary organic aerosols. *Environmental Science and Technology*, 43(24), pp. 9103–09.

Mudu, P., Martuzzi, M. and Alm, S. 2006. *Health effects and risks of transport systems: the hearts project.* Geneva: World Health Organization, pp. 97.

Peterson, J.E. and Stewart, R.D. 1970. Absorption and elimination of carbon monoxide by inactive young men. *Archives of Environmental Health*, 21, pp. 165–71.

Peterson, J.E. and Stewart, R.D. 1975. Predicting the carboxyhemoglobin levels resulting from carbon monoxide exposures. *Journal of Applied Physiology*, 39, pp. 633–8.

Salvi, S. 2007. Health effects of ambient air pollution in children. *Paediatric Respiratory Reviews*, 8(4), pp. 275–80.

Shima, M., Nitta, Y., Ando, M. and Adachi, M. 2002. Effects of air pollution on the prevalence and incidence of asthma in children. *Archives of Environmental Health*, 57(6), pp. 529–35.

Tang, D., Li, T.Y., Chow, J.C., Kulkarni, S.U., Watson, J.G., Ho, S.S., Quan, Z.Y., Qu, L.R. and Perera, F. 2014. Air pollution effects on fetal and child development: a cohort comparison in China. *Environmental Pollution*, 185, pp. 90–6.

Tymko, M. and Collins, D. 2014. Smoking bans for private vehicles: children's rights and children's voices. *Children's Geographies*, Nov, pp. 1–15.

Weichenthal, S., Dufresne, A. and Infante-Rivard, C. 2006. Indoor ultrafine particles and childhood asthma: exploring a potential public health concern. *Indoor Air*, 17(2), pp. 81–91.

World Health Organization (WHO) 2015. *Children's environmental health.* Available at www.who.int/ceh/risks/cehair/en/ (accessed 30 April 2016).

Zhang, Q. and Zhu, Y. 2010. Measurements of ultrafine particles and other vehicular pollutants inside schoolbuses in South Texas. *Atmospheric Environment*, 44(2), pp. 253–61.

Zhang, Q. and Zhu, Y. 2011. Performance of school bus retrofit systems: ultrafine particles and other vehicular pollutants. *Environmental Science and Technology*, 45(15), pp. 6475–82.

5   Is society now being social?

Helen Woolley

Public open spaces are the glue that holds the urban environment together. They connect buildings to each other and support travel between home, school, food collection, play, recreation and leisure activities (Woolley, 2003). The public open spaces that children use, or are expected to use, are predominantly parks, playgrounds, school playgrounds and streets. These networks of open spaces contribute to children's physical, mental and social health (World Health Organization, 1948) by supporting play, walking, cycling, meeting and other activities. However, disabled children are often under-represented amongst the users of such open spaces (Greenhalgh and Worpole, 1995; Swanwick *et al.*, 2002). In contrast, the issue of the provision for and wellbeing of children, usually understood as non-disabled children, in public open spaces has received relatively high levels of attention during the last 10–15 years (e.g. Holt, 2010; Hörschelmann and van Blerk, 2012). Such discourse has often focused on the increasing levels of obesity and reduction in children's independent mobility. Disabled children have not often been part of these mainstream discussions of wellbeing. This chapter therefore reminds the reader that the wellbeing of disabled children and the importance of their access to open space needs to be championed to a similar extent as that of non-disabled children.

Responding to this call reveals aspects of the disconnection that exists between policy, social attitudes and the wellbeing of disabled children. Although policy discourses emphasise inclusiveness, disabled people often report a range of social and physical barriers to their use of open spaces. These barriers include, but are not limited to, a lack of information about where accessible open space amenities are located; difficulty reaching an open space due to distance; lack of public transport and/or needing or wanting to be accompanied; difficulty accessing amenities such as toilets, rest places, shops and cafes; and being worried about safety and the cost of a facility (Kitchen and Law, 2001; Price and Stoneham, 2001). Other issues that may prevent the full inclusion of disabled children in open spaces include inadequate design of play and other spaces, and physical, social and/or organisational constraints (Woolley *et al.*, 2005). Moreover, disabled children are rendered powerless by their exclusion from decision making and design processes relating to the open spaces destined to meet their requirements and aspirations (Kitchen and Law, 2001).

To further explore some of the issues involved in the inclusion of disabled children in outdoor environments, this chapter draws upon the notion of 'Now Being Social' (Woolley, 2012). Now Being Social was first suggested as a way to succinctly explain the major discourses about play, childhood and disability in the context of seeking to understand some of the barriers to the provision of outdoor play spaces for disabled children. In this chapter the notion of Now Being Social will be briefly explained before the discussion moves on to explore some ways in which this concept can be understood in practice and daily life with respect to outdoor environments that disabled children might use.

Disability: medical or *social* model?

Society (in the global North) often seems to have a limited understanding of disabled people, including children, with a dominant notion being that most disabled people and children are wheelchair users (Mathers, 2008). This is also the case in some research where the main focus has been on wheelchair users rather than on people with a range of impairments (see, for example, Kitchen and Law, 2001). A deeper understanding of the range of impairments that people might experience can be gained from the law, research and policy guidance emanating from government, charities or other agencies that indicate the need to consider a diversity of impairments in order to understand and foster disabled children's wellbeing.

Impairments and the nomenclature used to describe them take various forms. The code of practice for special educational needs suggests four broad areas of need: communication and interaction; cognition and learning; social, emotional and mental health difficulties; sensory and/or physical needs (Department for Education and Department of Health, 2015). Alternative terminology is used in research and guidance for the provision of grounds of special schools (schools that are exclusively for disabled children) including mild, moderate and severe learning difficulties; autism; blind and visual impairments; deaf and hearing impairments; multi-sensory impairments; emotional and behavioural difficulties; and language disorders (Stoneham, 1996). Later research about the inclusion of disabled children in mainstream primary school playgrounds included children with autism, spina bifida, cerebral palsy, muscular dystrophy, visual impairments, hearing impairments and communication difficulties (Woolley *et al.*, 2005). Medical research has highlighted children with sight, speech and hearing impairments but also children who need support in the administration of medicine, assisted feeding and the transport of bulky equipment such as oxygen cylinders, ventilators and suction equipment (Hewitt-Taylor, 2009). More recently understandings of impairment have broadened to consider behavioural differences, issues of mental health, the implications of 'contested illnesses such as fibromyalgia and chronic fatigue syndrome' and the possible impacts of 'disabling and marginalising aspects of fatness' (Chouinard, 2010: 242–3).

Collectively these pieces of work, whether derived from law, guidance or research, provide society with the possibility of an enhanced understanding of the breadth of impairments reflecting mind–body–emotional differences (Holt, 2007).

Disability should therefore be more explicitly and widely understood in the context of the World Health Organization's definition of health as being 'a state of complete physical, mental and social wellbeing and not merely the absence of disease or infirmity' (World Health Organization, n.d.).

Much of the practice and earlier research that has informed these definitions and understandings about disabled children and their play in both indoor and outdoor environments has been undertaken by physiotherapists, medical clinicians and educationalists, not built environment researchers or practitioners who plan, design and manage outdoor environments. In the past, a medical model of disability, in which impairment was understood to be a personal tragedy in need of individual treatment, was influential (Oliver, 1985). Forty years ago this medical model of disability was challenged by activists and then increasingly by specific academics, resulting in the social model of disability, which was first suggested by the Union of the Physically Impaired Against Segregation (UPIAS) in the United Kingdom (UPIAS, 1976). In contrast to the medical model of disability, the social model of disability has an underlying assumption that society disables people through the barriers they face within social structures and institutions (Oliver, 1985) and by policies and practices (Barnes, 1991; Zarb, 1995). Rather than disability being a personal tragedy, disability is therefore understood as a social construction. As Gleeson (1999) observes, within the social model, social and spatial processes can disable rather than enable people with impairments. To encourage and foster disabled children's wellbeing we should foster more explicitly enabling social and spatial processes and disabled children's rights to participate in society.

At international and national levels various rules and legal directives exist as frameworks for asserting rights for disabled people. The United Nations Standard Rules on the Equalisation of Opportunities for Persons with Disabilities (United Nations, 1993) enshrines at an international level the right to access the physical environment for *all* people. This is then expressed in different countries through a range of mechanisms including law, policy and practice. For example, in England the Disability Discrimination Act 1995 confirms that it is unlawful to treat disabled people less favourably than non-disabled people. The definition of a disabled person was enhanced in the Equality Act 2010 (Department for Culture, Media and Sport (DCMS), 2010: 4) and includes people who have a physical or mental impairment that has 'a substantial and long-term adverse impact on a person's ability to carry out normal day-to-day activities'. Service and facility providers are required to make 'reasonable adjustments' in order that disabled people can access services. On the surface, it would seem that these international and national rules might support the social model of disability but this depends upon the extent to which such rules are realistically put into practice and support the daily living of disabled people.

Changing childhoods: becoming an adult or *being* a child

It is not only the understanding of disability that has changed over the years. Parallel changes have occurred around the notion of childhood. Discourses of disability have shifted from a medical to social model, and in a similar timeframe the notion of childhood as a social construct has emerged to challenge the earlier framing of childhood as being biologically determined (James and Prout, 1997). This social construct suggests that children are not 'becoming' adults but that they are 'being' individuals in their own right in the current time. A range of social variables, such as their class, gender, ethnicity and culture, contribute to any child's state of being and can result in multiple identities for an individual child. In addition, impairment/s can contribute to a child's identity and state of being (James and Prout, 1997).

The previous discussion about children with special needs, as expressed in English society in the Special Educational Needs Act, is an example of children's needs being a social construct that can mask both uncertainty and disagreement about what might be in the best interests of a child (Woodhead, 1997). Woodhead suggests that the term 'needs' carries assumptions of the cultural and personal perspectives of the person using the term, not of the child, and thus can result in the term being understood differently by different users or hearers of the term. In contrast to the language of needs, Woodhead suggests an approach using the notion of children's rights. For children, this is expressed at the international level in the United Nations Convention of the Rights of the Child (United Nations, 1989). This sets an international framework for children as beings in the here and now, not as people who are becoming adults. In some parts of the world, children's rights are expressed through the concept of Child Friendly Cities, initiated by UNICEF and which suggests a range of building blocks that can be used to take policy and practice forward for a range of areas of life. Possibly the best expression of children's rights in the UK was under the Labour government of 1997–2010 where these rights were reflected in a range of English policies, including Every Child Matters (Chief Secretary to the Treasury, 2003), a 10-year Children's Plan (Department for Children, Schools and Families (DCSF), 2007) and a Play Strategy (DCSF and DCMS, 2008). These disappeared with the Conservative/Liberal Democrat coalition that took power in 2010 when policies about children were increasingly focused on education and related tests and examinations. The concept, expressed in policy, of children as beings in their own right slipped off the agenda and was not mentioned in the general election campaign of 2015 and is not a priority for the current Conservative government.

Play: fun and benefits for *now*

Play is something that both adults and children do, but is often considered to be the domain of children. Indeed, play has been suggested as part of the nature and culture of childhood (James *et al.*, 1998) and as being something children do anywhere (Opies, 1969) and anytime. Many people from different disciplines and backgrounds have tried to understand, define and categorise play, with Huizinga (1938) suggesting that play might be a constituent part of human civilisation. The complex and fluid nature of play and the lack of agreement about what play is is explored in the seminal work of Sutton-Smith (1997) who discusses the ambiguity of play. Other literature discusses what play can do for individuals, families and society and conversely what the lack of play can mean (Cole-Hamilton *et al.*, 2002; Lester and Russell, 2008). A more recent discussion reflects back to the broader concept that play is not only for children, but that it happens throughout life and is 'a form of coming to consciousness and a way to be "otherwise" and relates to the cultivation of ethical generosity' (Woodyer, 2012: 322). Much of the discourse about children's play has positioned play as 'a mechanism for learning and development' (Lester and Russell, 2008: 13) embracing the view that children's play is part of the preparation for becoming an adult and in this way links with the biological discourse about childhood. However, play has intrinsic value for children, apart from what it might do for other people, families and society and for development into adulthood – play is important for children in the here and now.

The importance of children's play is captured at the international level in the United Nations Convention on the Rights of the Child in which Article 31 asserts that 'State Parties recognize the right of the child to rest and leisure, to engage in play and recreational activities . . . and to participate freely in cultural life and the arts' (United Nations, 1989: 10). Early in 2013, the United Nations released a General Comment reaffirming the importance of Article 31, which was considered by some to have become 'the forgotten article' (International Play Association, 2013: 1). In England, the child's right to play has not been continually expressed in policy, but was for a short period of time between 2008 and 2011 when the Labour government developed a Play Strategy (DCSF and DCMS, 2008). This was preceded by National Lottery ring-fenced funding for play, while the Play Strategy (DCSF and DCMS, 2008) was accompanied by a funding programme for the refurbishment of existing and the development of new play spaces across England. These policies and programmes sought to take an inclusive approach to provide play spaces that could be used by disabled and non-disabled children.

Now being social

It is from these discourses of disability, childhood and play that the concept of Now Being Social (Woolley, 2012) developed. The concept can be understood in the following way:

> Now: play is not (just) about preparing to be an adult but has meaning for children *now*;
>
> Being: childhood is not (just) about the biological issue of becoming an adult but is also about *being* a person now;
>
> Social: being disabled is about being a person and the barriers that restrict full participation in society – the *social* rather than medial model of disability.
>
> Woolley, 2012: 3

In each discourse the inclusion of '(just)' is made because the theories informing Now Being Social are drawn from research and, to some extent, practice but many individuals and individual children do not experience their daily lives within the framework of Now Being Social; instead, for many children play is directed by others and towards educational or future benefits. Play is not a social experience for these children, just as being disabled is still about a medical condition or state of being for many disabled children.

One way that Now Being Social might be expressed in society is in an inclusive approach to disabled children and the outdoor environment, whether that be the housing areas, parks, playgrounds, school playgrounds or streets suggested in the introduction as being the types of open spaces that children might use most in their daily lives. For many years, separation or exclusion of disabled children from non-disabled children has characterised outdoor play environments, a situation variously called segregation (Wheway and Millward, 1997) or apartheid (Gleeson, 2001). Despite, or possibly because of, a general lack of a common understanding of what inclusive design might mean, the Disability Rights Commission in the UK, for example, suggested six principles for inclusive design: ease of use; freedom of choice and access to mainstream activities; diversity and difference; legibility and predictability; quality; and safety (Goodridge, 2008). Elsewhere, three dimensions to an inclusive approach are suggested, which draw upon a range of research and practice. The three dimensions are that disabled children should be able to play in an open space; disabled and non-disabled children should be able to play together in the open space; and families should also be able to use the open space (Woolley, 2012).

Theory into practice?

This section draws upon two studies that have sought to understand more about the provision of neighbourhood open spaces that seek to be inclusive for disabled children. The first study explored issues of inclusion of disabled children in the playgrounds of mainstream primary schools, and the second one relates to the provision of an inclusive facility in a special school.

Experiences in the playground of a mainstream school

Research in six mainstream schools in Yorkshire in the north of England included disabled children as participants. These children were often actors in their own right, asserting their right to play and sometimes challenging the social and organisational practice within their schools (Woolley *et al.*, 2006). This research identified good and poor examples of physical, social and organisational practices for enabling disabled children to play within the school playground. One noteworthy aspect was that good practice in the physical environment supporting disabled children was predominantly confined to the indoor spaces, taking the form of larger computer work stations, disabled toilets and ramps for access into and out of the buildings. No other real consideration had been given to physical issues in the outdoor playgrounds, resulting in poor physical practice in most of the schools.

Despite the lack of physical good practice there was some experience of good social and organisational practice to support the inclusion of disabled children in both the initiation of play and ongoing play on school playgrounds. Good organisational practice was identified for one child when her physiotherapist realised the importance of play and lunch times, and arranged for the child's physiotherapy to take place during assembly or quiet reading time, not in play time. In another situation, the organisational structure allowed for a child's personal assistant, who understood the child's impairment, to take them for unscheduled outside breaks if they felt the child would particularly benefit from this additional outdoor experience at a specific point in time.

Good social practice could be the result of action by staff or the agency of the disabled child or their friends. In considering the inclusion of the disabled children in play on the playgrounds, three strategies were identified as being used by the disabled and non-disabled children. These were the actions of the disabled children themselves; the use of the disabled children's equipment; and the adaptation of play by friends of the disabled children. For example, one of the disabled children not only fully engaged in the play but led it by acting as a computer that beeped to tell her friends there was a mission to undertake. Games of 'tiggy' mean that one child is 'on' and seeks to catch another child who is then 'on'. In order not to be caught, a child is safe if they do something specific such as being off ground or by touching a specified colour. Participation in such games by one disabled child was facilitated by her walker on which she could sit and be safe in 'tiggy off ground', or she could touch when blue was called in 'tiggy colour' because the blue on part of her walker provided her with an immediate refuge when the colour blue was called. The importance of friends enabling the inclusion of disabled children was revealed in different situations. One of the most visually vivid of these examples related to play on a piece of fixed play equipment in the form of a low wooden bridge that was suspended just above the floor level between two vertical posts. The children walked and jumped on this bridge and in order to allow the disabled child to be included the friends would help the disabled girl onto the bridge and then jump on the bridge to gently bounce it up and down resulting in clear expressions of joy and delight by the disabled child. In these examples, the disabled children sometimes initiated actions in the playground whilst in other situations friends supported the individual disabled child.

Sometimes the safety concerns of adults associated with a disabled child could have an impact on the inclusion of a disabled child in play activities. One example of this was where a disabled child wanted to use a piece of equipment made from tyres. When there was no member of staff present the child was observed playing with other children on this equipment. However, the child's Personal Support Assistant (PSA) considered that the child was not able to use the equipment without hurting herself and actively prevented the child from using the equipment. Another child told of her dissatisfaction that she could not do everything her peers could do. She was particularly keen to roll down the grass slope, first going up the steep rather than the less steep side. The PSA encouraged the child while taking into account a previous fall on the slope. The child's determination meant that she was able to succeed in her desire to roll down the slope, and as she completed the roll her peers cheered and she was left with a great sense of achievement. This was a self-affirming experience of risk-taking driven by the agency of the child.

Experiences in the playground of a special school

Although acknowledging that the exclusion of disabled children from some play spaces is a reflection of their experiences in other types of settings, Jeanes and Magee (2012) explore theory, policy and practice with respect to inclusive outdoor play. In this instance the understanding of an inclusive facility was focused around the physical issues of inclusive play equipment, provision of toilets, ease of access and suitable surfacing, together with social issues of positive experiences with peers and the role of staff. The desire for an inclusive play space was driven by the parents of children at a special school who were concerned about a lack of inclusive play facilities in a city in the UK. Parents, supported by the head teacher, worked with the city council to attract funding from different sources for a new outdoor play space and staff to facilitate the use of the facility. The aim was for the facility to be inclusive, fun, enjoyable and safe and to this end it included pieces of adaptive equipment that could be used by both disabled and non-disabled children. The facility was available to schools during the day and the public at evenings and weekends when play workers staffed the facility. It is of interest to note that the facility was developed at the special school, rather than in an alternative neighbourhood open space such as a park.

Many of the children's and parental responses were very positive, with the experience of inclusion attributed to four factors: the location of the facility, the respondents' involvement in the design, the absence of discriminatory attitudes and the presence of play workers. Being located at the special needs school helped create a culture of inclusion and acceptance for both the disabled children and their families who also commented that they felt safe. There was a feeling that the disabled children belonged there, possibly because of an established culture of tolerance and acceptance of difference on the established special school site. Disabled children and their parents were involved in the design process and influenced the design of the outdoor play space. Some families remained involved in the design process into the management of the facility, which increased their sense of connection and identification with the outdoor play space.

Also contributing to the inclusive nature of the facility was the absence of discriminatory attitudes, which appeared to be the result of two main factors. First, being at the special school, the disabled children were familiar with the location and existing ethos of the site: a reversal of the often usual power dominance of able-bodied people (Imrie, 1996). Second, the disabled children and their parents who had been involved in the design process for the facility had a sense of ownership of the facility. This also seemed to contribute to the absence of discriminatory attitudes. Both of these approaches could have resulted in non-disabled children not wanting to use the facility, but this was not the case. The research team observed that half the users were non-disabled children. Explanations for this outcome included the efforts made by the families of the disabled children who drew upon their own experiences of exclusion to encourage the inclusion of the non-disabled children. Another factor contributing to the inclusive nature of the facility was the presence of trained play workers who supported the play of all children at the facility. By not focusing their attention exclusively on the disabled children, the play workers supported non-disabled children's positive perceptions towards the disabled children. Some parents particularly discussed that this was very different from their previous experiences of leisure facilities.

Conclusion

In this chapter a range of rights, policies and social responsibilities relating to the play opportunities of disabled children have been discussed. Rights relate to those of the dual identities of childhood and disability and are expressed at the international level through United Nations policies and conventions. These rights cascade into policy at a national level, although such policies are not impervious to political changes, such as happened in the UK after recent general elections. Policy is therefore put into practice through various societal mechanisms. To illustrate this process, the chapter draws on endeavours to support disabled children's play made by two societal institutions – mainstream schools and a special school.

From the research in mainstream schools, it can be seen that there are ways in which disabled children can be included in play on traditional primary school playgrounds. The enjoyment that such play can give was evident where disabled children themselves were agents of their inclusion in play, where peers facilitated the inclusion of disabled children in play, and where staff supported the sometimes risky play that a disabled child wanted to engage in. At such times the children were playing in the here and now and not driven by any future benefit for their education or development. Sometimes a child asserted their agency by insisting on undertaking a play activity that might have been thought not possible or too risky by others: a very real expression of a children's agency.

In the special school context, the play facility aimed to be fun and enjoyable for all children. The involvement of disabled children and their parents in the design of the facility provided some level of power for the children and their families and resulted in a strong sense of ownership of the outdoor play space. Furthermore, the presence of play workers who engaged with all children – disabled and non-disabled – clearly contributed to an expression of the social model of disability.

When rights, policy and social organisations come together to support disabled children's play in these contexts it appears that specific aspects of Now Being Social can be realised in the daily lives of disabled children, adding to their sense of wellbeing. But there were also examples in which practice was not constructed within this framework. One of the most noticeable examples of an absence of inclusive practice was the lack of consideration given to the physical outdoor environment at the mainstream schools where, despite laws being in place, no real provision had been made to support the disabled children's play.

For a more appropriate approach, the multiple rights of disabled children to play need to be translated into law and/or policy and then put into practice in the contexts of such schools and other areas of daily life. This cascading of rights to policy to practice should have a positive impact on the daily lives of disabled children, allowing them to play in the here and *now, being* a child in their own way and contributing to their wellbeing, with society taking responsibility through the *social* model of disability to provide an enabling external environment.

References

Barnes, C. 1991. *Disabled people in Britain and discrimination: a case for anti-discrimination legislation*. London: Hurst/British Council of Organisations of Disabled People.

Chief Secretary to the Treasury. 2003. *Every child matters*. London: The Stationery Office.

Chouinard, V. 2010. Impairment and disability. In: V. Chouinard, E. Hall, and R. Wilton, eds. *Towards enabling geographies: 'disabled' bodies and minds in society and space*. Farnham: Ashgate, pp. 241–57.

Cole-Hamilton, I., Harrop, A. and Street, C. 2002. *Making the case for play: gathering the evidence*. London: National Children's Bureau.

Department for Children, Schools and Families (DCSF). 2007. *The children's plan: building brighter futures*. London: DCSF.

Department for Children, Schools and Families (DCSF) and Department for Culture, Media and Sport (DCMS). 2008. *The play strategy: a commitment from the children's plan*. London: DCSF.

Department for Culture, Media and Sport (DCMS). 2010. *Equality Act 2010*. London: The Stationery Office. Available at www.legislation.gov.uk/ukpga/2010/15/contents (accessed 27 February 2017).

Department for Education and Department of Health. 2015. *Special educational needs and disability code of practice: 0 to 25 years*. London: Department for Education. Available at https://www.gov.uk/government/uploads/system/uploads/attachment_data/file/398815/SEND_Code_of_Practice_January_2015.pdf (accessed 28 February 2017).

Disability Discrimination Act. 1995. Available at www.legislation.gov.uk/ukpga/1995/50/contents (accessed 31 March 2011).

Gleeson, B. 1999. *Geographies of disability*. London: Routledge.

Gleeson, B. 2001. Disability and the open city. *Urban Studies*, 38, pp. 251–65.

Goodridge, C. 2008. *Inclusion by design: a guide to creating accessible play and childcare environments*. London: Kids, Playwork Inclusion Project with Department for Children, Schools and Families.

Greenhalgh, L. and Worpole, K. 1995. *Park life*. Bourne Green: Comedia Demos.

Holt, L. 2007. Children's sociospatial (re)production of disability within primary school playgrounds. *Environment and Planning B*, 25, pp. 783–802.

Hewitt-Taylor, J. 2009. Children with complex, continuing health needs and access to facilities. *Nursing Standard*, 23, pp. 35–41.

Huizinga, J. 1938. *Homo Ludens: a study of the play element in culture*. Boston, MA: Beacon Press.

Kitchen, R. and Law, R. 2001. The socio-spatial construction of (in)accessible public toilets. *Urban Studies*, 38(2), pp. 287–98.

Imrie, R. 1996. *Disability and the city: international perspectives*. London: Paul Chapman.

82 *Helen Woolley*

International Play Association (IPA). 2013. UN stands up for children's right to play, arts and leisure in a landmark moment for children. IPA Press Release, 1 February 2013. Available at http://ipaworld.org/childs-right-to-play/un-general-comment/un-stands-up-for-childrens-right-to-play-arts-and-leisure-in-a-landmark-moment-for-children/ (accessed 28 February 2017).

James, A. and Prout, A. eds. 1997. *Constructing and reconstructing childhood: contemporary issues in the sociological study of childhood*. London: Routledge.

James, A., Jenks, C. and Prout, A. 1998. *Theorizing childhood*. Cambridge: Polity Press.

Jeanes, R. and Magee, J. 2012. Can we play on the swings and roundabouts? Creating inclusive play spaces for disabled young people and their families. *Leisure Studies*, 31(2), pp. 193–210.

Holt, L. 2010. Young people's embodied social capital and performing disability. *Children's Geographies*, 8, pp. 9–21.

Hörschelmann, K. and van Blerk, L. 2012. *Children, youth and the city*. London: Routledge.

Lester, S. and Russell, W. 2008. *Play for a change*. London: Play England.

Mathers, A. R. 2008. Hidden voices: the participation of people with learning disabilities in the experience of public open space. *Local Environment*, 13(6), pp. 515–29.

Oliver, M. 1985. The integration segregation debate: some sociological considerations. *British Journal of the Sociology of Education*, 6(1), pp. 75–92.

Opie, I. and Opie, P. 1969. *Children's games in street and playground: chasing, catching, seeking, hunting, racing, duelling, exerting, daring, guessing, acting, pretending*. Oxford: Clarendon Press.

Price, R. and Stoneham, J. 2001. *Making connections: a guide to accessible greenspace*. Bath: The Sensory Trust.

Stoneham, J. 1996. *Grounds for sharing: a guide to developing special school sites*. Winchester: Learning through Landscapes.

Sutton-Smith, B. 1997. *The ambiguity of play*. Cambridge, MA: Harvard University Press.

Swanwick, C., Dunnett, N. and Woolley, H. 2002 *Improving urban parks, play areas and green spaces*. London: Office of the Deputy Prime Minister.

Union of the Physically Impaired Against Segregation (UPIAS). 1976. *Fundamental principles of disability*. London: The Disability Alliance.

United Nations. 1989. *The United Nations Convention on the Rights of the Child*. New York, NY: United Nations. Available at http://353ld710iigr2n4po7k4kgvv-wpengine.netdna-ssl.com/wp-content/uploads/2010/05/UNCRC_united_nations_convention_on_the_rights_of_the_child.pdf (accessed 28 February 2017).

United Nations. 1993. *The United Nations Standard Rules on the Equalisation Opportunities for Persons with Disabilities*. New York: United Nations.

Wheway, R. and Millward A. 1997. *Child's play: facilitating play on housing estates*. Coventry: Chartered Institute of Housing with support from the Joseph Rowntree Foundation.

Woodhead, M. 1997. Psychology and the cultural construction of children's needs. In: A. James and A. Prout, eds. 1997. *Constructing and reconstructing childhood: contemporary issues in the sociological study of childhood*. London: Routledge. pp. 63–77.

Woodyer, T. 2012. Ludic geographies: not merely child's play. *Geography Compass*, 6(6), pp. 313–26.

Woolley, H. 2003. *Urban open spaces*. London: Spon Press.

Woolley, H. 2012. Now Being Social: the barrier of designing outdoor play spaces for disabled children. *Children and Society*, 27 (6), pp. 448–58.

Woolley, H., Armitage, M., Bishop, J., Curtis, M. and Ginsborg, J. 2005. *Inclusion of disabled children in primary school playgrounds*. London: National Children's Bureau and Joseph Rowntree Foundation.

Woolley, H., Armitage, M., Bishop, J., Curtis, M. and Ginsborg, J. 2006. Going outside together: good practice with respect to the inclusion of disabled children in primary school playgrounds. *Children's Geographies*, 4(3), pp. 303–18.

World Health Organization (WHO). 1948. Preamble to the Constitution of the World Health Organization as adopted by the International Health Conference, New York, 19 June–22 July 1946; signed on 22 July 1946 by the representatives of 61 states (*Official Records of the World Health Organization*, 2, 100) and entered into force on 7 April 1948. The definition has not been amended since 1948.

World Health Organization (WHO). n.d. Constitution of WHO principles. Available at www.who.int/about/mission/en/ (accessed 28 February 2017).

Zarb, G. ed. 1995. *Removing disabling barriers*. London: Policy Studies Institute.

6 Cities that encourage happiness

Claire Freeman

Can cities be happy places, and what are the features that can make cities happy places for those who live there? Consideration of this question in turn generates more questions: what is happiness, does happiness matter and, in relation to this book, can cities be happy and indeed healthy places for children? In this chapter I grapple with these questions and put forward the case that happiness is neither a random occurrence nor a trait purely related to an individual's temperament. Rather, happiness is a state of being that can be enhanced and supported by the nature and design of the cities that people live in, and in the lives that the city encourages them to lead. Happiness is perhaps the strongest determinant of health and wellbeing, specifically mental wellbeing, as well as being an outcome of good health. Children have a propensity towards happiness that is supported or thwarted by the physical, social and emotional conditions in which they live. In the chapter I explore what it is that helps engender 'happiness' in children in our built environments, and conclude with a section focusing on 'fun' and happiness in cities, showcasing just some of the ways that my own city – Dunedin in the South Island of New Zealand – creates happy, fun spaces, places and happenings for its children.

The catalyst for this chapter comes from research conducted between 2006 and 2014, with children in a number of different neighbourhoods in New Zealand and Fiji (Freeman, 2010; Freeman *et al.*, 2015). As a member of a team conducting research in schools in New Zealand, Fiji and Kiribati, we spent a good deal of time talking to children aged 9–12 about where they lived, their schools, neighbourhoods and their lives generally. It became clear when we talked to children in their schools during our research, either during formal interviews, chatting with the children between interviews or when having a tea break in the playground, that some groups of children seemed to radiate happiness when talking about their lives, the neighbourhoods they lived in and the people they knew. This expression of what we call 'happy feel' came from children who were very interactive between themselves and the adults, seemed confident, chatty, often physically connecting (holding hands, playful pushing), and radiated energy and eagerness as a group. We experienced this feeling in schools situated in New Zealand neighbourhoods that on the surface appeared to have little in common. Of the two schools where this happy feeling was particularly apparent, one was at the lowest end of the socio-economic spectrum, had high deprivation rates and was on a cold hillside that bears the full brunt of Antarctic winds. It was in a state housing area that frequently appears in the local papers in association with negative stories. The other school was the reverse, located in a high socio-economic suburb, with an idyllic harbour-side setting with its own sandy beach and known for its sunny aspect.

Later research (Freeman *et al.*, 2015) took us to Fiji where we encountered a similar feeling to that first school we visited – a low income school for indigenous Fijians again had children who exuded a happy feeling, as did one of the village schools we attended. This happy feeling was not so apparent in some of the other schools we visited. These were usually more middle income or academically higher achieving schools. Individually, though, no matter which schools we visited in both countries, the children were bright, keen and almost unfailingly enthusiastic (we met very few clearly unhappy children), but this was happiness as individuals rather than as groups.

So what is it that creates this palpable sense of happiness in some groups of children in some places? Happiness is clearly infectious, as indicated by 'a group happiness aura' and results from a set of circumstances apparent in some of the places where children gather and live.

Happiness has a spatial component and has been a focus of increasing attention for health geographers who are particularly interested in the spatial dimensions of wellbeing (Kearns and Collins, 2009; Conradson, 2012). As a starting point for analysis of space and wellbeing and, in turn, happiness, Fleuret and Atkinson (2007) identify three spatial elements whose presence can be seen to contribute to what I refer to as 'tangible feelings of happiness' or an 'aura of happiness'. These elements are integrative spaces; secure spaces and therapeutic spaces, examples of which are given later in this chapter. Concurrent with this developing interest in wellbeing and place has been the burgeoning interest in emotional geographies and how places shape feelings of contentment and happiness (Kearns and Collins, 2009). Assessments through the lens of children's emotional geographies view wellbeing as inseparable from the social, cultural, economic and political landscapes of childhood (Blazek and Windram-Geddes, 2013). In this chapter I explore what the circumstances and the spatial components of city landscapes might be that contribute to both happiness and wellbeing, and consider how we as professionals working in urban environments might follow Bhutan's lead with its 'gross national happiness index' (Centre for Bhutan Studies & GNH Research, 2015) and design cities with happiness creation for its citizens as a core value.

Why happiness?

The word 'happiness' is commonly used, easily recognisable, but yet can be hard to define. It fails to conform to a set of standard criteria, and yet it is a state of being that, like its converse 'misery', is relatively easy to discern. Although happiness can be to some extent culturally determined, a general definition regardless of culture can be formulated as follows: 'Happiness is defined as a positive emotional state that is most general [but] . . . this experience is embedded in specific socio-cultural contexts and circumstances' (Uchida *et al.*, 2004: 226). Thus, Uchida *et al.* identify a relationship between happiness and the contributing environment. The *Journal of Happiness Studies*, which is the primary outlet for 'happiness' research, does not specifically define happiness but does identify an overarching context for happiness that they refer to as 'subjective wellbeing'. This journal encourages papers focusing on happiness, including life-satisfaction and the affective enjoyment of life, which they see as constituting states of happiness. Furthermore, the journal expresses a specific interest in 'health-related quality-of-life research'.

The relation between happiness and health is inferred, if not specifically stated, in much of the research explored in this chapter. The primary focus in this research tends to be on psychological/subjective aspects of wellbeing rather than physical health (Proctor *et al.*, 2009; Uusitalo-Malmivaara and Lehto, 2013). The need for more work that directly explores the relationship between subjective wellbeing or happiness and physical health and illness has been strongly supported by researchers interested in happiness studies (Uchida *et al.*, 2004). This research deficiency is somewhat ironic given the well-researched and established relationship between poor health and unhappiness. In this respect Graham *et al.* (2011) go so far as to identify which health conditions in different countries cause greatest unhappiness or 'life satisfaction'. In a fascinating study of how public policy can be used to 'minimise misery' Lelkes (2013) found that 'individuals with health impairment are less likely to be very happy and more likely to be unhappy [with] the size of the effect for unhappiness being greater' (2013: 131). They conclude by pointing to the need 'for early interventions to promote children's emotional well-being' (2013: 135), going on to state that with children such interventions are likely to be very effective. Others note the relationship between poor mental health and certain types of neighbourhood characteristics (Pearson *et al.*, 2014), suggesting that place does influence wellbeing. The latter part of this chapter reviews interventions with respect to the built environment, but first it is necessary to explore further the happiness concept.

The happiness concept

What then are the generic constituents of happiness? At its most fundamental level, happiness is dependent on the presence of certain basic provisions in life, namely food, access to health, shelter, education, freedom and safety, and other such essentials as recognised in the 1948 Universal Declaration of Human Rights. These rights include the right to a standard of living adequate for the health and wellbeing. For children, these fundamental rights were further enshrined in the United Nations Convention on the Rights of the Child 1989 (United Nations, n.d.). A state of happiness thus requires access to basic needs and respect for people's rights, a finding supported by a study of neighbourhood satisfaction in Uruguay (Gandelman *et al.*, 2012). This study found, not unsurprisingly, that access to public goods such as water, electricity, health, a sewage and waste disposal system, street trees, sidewalks, an absence of pollution, together with social life, family, work and leisure were necessary attributes for life satisfaction. Assuming these basic necessities are in place, what are the next building blocks for happiness?

Surprisingly, economic advance does not necessarily equate to enhanced happiness. Studies in Norway (Hellevik, 2003) found that the increasing tendency to prioritise income and material possessions seems to reduce happiness. Similarly, in China Brockmann *et al.* (2009: 387) noted that between 1990 and 2000 happiness 'plummeted despite massive improvements in living standards'. Another Chinese study found that elderly Chinese who live with their grandchildren have higher levels of happiness than their counterparts (Tafarodi *et al.*, 2012). Interestingly, and counter to common belief, increased pressure and a faster pace of life does not necessarily reduce life satisfaction and happiness (Garhammer, 2002), a finding that Garhammer relates to the fact that the 'most active people are typically the most happy' (2002: 249). This is an observation we have noticed when interviewing children in our research (Freeman, 2010; Freeman *et al.*, 2016). Although we expected children with very busy lives, full of activities and often going between multiple carers and locations to be stressed and unhappy, they generally seemed to take what seemed to us to be a 'hectic' life in their stride. Indeed, they often spoke with pride about their numerous activities. Furthermore, they also usually seemed to have good social relations with their peers.

A four nation study of university students on 'what makes for a good life?' found variation, but also consistency between nations, with a good life being prominently associated with social concerns, mainly relationships with family and friends (Tafarodi *et al.*, 2012). Family health was identified as being especially important. The authors noted that this is an interesting finding in light of the 'global popular culture that promotes self-expression and consumerist individualism' (2012: 798). Similar findings on the importance of social relations emerge from other studies that point to the relation between happiness and social harmony (Uchida *et al.*, 2004) and friendships. For children, 'good relationships with family strongly predicted high happiness' and 'for both genders a small number of friends predicted low happiness' (Demir and Weitekamp, 2007: 611). Social connectedness as in friends and family repeatedly emerges as a key factor for children in happiness and good mental health across a varied range of localities and contexts: the US (Csikszentmihalyi and Hunter, 2003; Chaplin, 2009), Finland (Uusitalo-Malmivaara and Lehto, 2013), India (Holder, 2012), Australia, the UK (Sargeant, 2010) and South Africa (Eloff, 2008). A final survey worthy of note in this context is one that asked adults to think back to their childhoods and recall what made for a happy childhood. It was found that memories recalling states of happiness were strongly associated with social events and social experiences. This finding led the authors to conclude that 'a happy childhood is defined primarily by the nature of interaction with others' (Batcho *et al.*, 2011: 541).

The two key determinants of happiness as identified in happiness studies are thus social connections and access to those resources necessary for a reasonable quality of life. Attention now turns to how happiness can be related to place, and how the positive attributes of space can be maximised in ways that enhance happiness in our built environments.

Happy places in the built environment

Although there is an increasingly substantial body of research on health, happiness and, to a lesser extent, children's happiness, very little of this research looks at how place and space impact on happiness. Beyond some rather perfunctory acknowledgement of the basic necessities required to meet children's fundamental physical and emotional needs, it is almost as if happiness and children's lives are lived in a spatial vacuum. In comparison to the weight of books on children and depression, mental illness, achievement and relationships, there is a gap in terms of publications on the character of 'happy' environments. Indeed, where there is information on cities and, to use the broader term, 'subjective wellbeing' there is a somewhat negative overtone, suggesting that cities are counter-productive in terms of happiness building, especially in poorer urban communities and especially for children (O'Campo *et al.*, 2009; Schaefer-McDaniel, 2009; Khan *et al.*, 2011; Anakwenze and Zuberi, 2013). Given the rising proportion of the population globally, including children living in the city, this is a matter for concern. Montgomery, author of an otherwise inspirational book *Happy City*, begins his concluding chapter with the following depressing words:

> No age in the history of cities has been so wealthy. Never before have our cities used so much land, energy and resources. . . . Despite all we have invested in this dispersed city, it has failed to maximise health and happiness. It is inherently dangerous. It makes us fatter, sicker, and more likely to die young. It makes it harder to connect with family, friends, and neighbours. It makes us vulnerable to the economic shocks and rising energy prices inevitable in our future. As a system, it has begun to endanger both the health of the planet and the well-being of our descendants.
>
> Montgomery, 2013: 316

Fortunately though, as Montgomery identifies elsewhere in his book, cities are more than economic, and even built structures can nonetheless be inhabited in ways that do reflect positive ways of being. This is not to deny the importance for children of economics. For example, children are far more likely to thrive if parents are in employment and are not overworked or underpaid. The built environment is important in providing access to good schools and safe streets. However, even where the economic and built environment is challenging, there will still be opportunities and environmental conditions that can be accentuated to enhance children's happiness (see Table 6.1 and Table 6.2 for examples). What is further encouraging is that these can often be easily and indeed cheaply achieved. Children have a propensity towards happiness that is supported or thwarted by the physical, social and emotional conditions in which they live. I now look at how these conditions can be enhanced through two primary city concepts: green cities and convivial cities.

Green cities

There is a well demonstrated connection for adults between natural spaces and wellbeing, especially mental wellbeing (Mayer *et al.*, 2004; Korpela and Yle'n, 2007; Mitchell and Popham, 2008; Ryan *et al.*, 2010; Howell, *et al.*, 2011; Lachowycza and Jones, 2013; MacKerron and Mourato, 2013). Contact with nature has also been shown to be vital to the health, happiness and wellbeing of children, both mentally and physically (Kahn and Kellert, 2002; Louv, 2005; Holbrook, 2008; England Marketing, 2009; McCurdy *et al.*, 2010). Nisbet and colleagues, in their paper engagingly entitled 'Happiness is in our nature', argue that not only is nature good for us in terms of enhancing positive mental wellbeing, but there are also benefits for nature because 'happy people are more likely to act environmentally' (Nisbet *et al.*, 2011: 318). As most children in developed countries live in urban areas, and the proportion of people in urban areas is expected to rise to 6 in every 10 people by 2050 (United Nations, 2014), then it is clear that the access to green spaces within the urban environment is required for children's happiness and it needs to be an integral part of city provision.

Louv (2005) puts forward a compelling argument for the risk to children's physical and mental health associated with a lack of access to and connection with nature. However, in most but not all cities in Europe, Australasia, Northern America and Japan, green spaces are present (Freeman and Tranter, 2011) in the form of parks, gardens, sports grounds, street trees, vacant land along roadsides and footpaths, landscaping around hospitals, businesses and schools, and private gardens. In such places, the issue is often not a lack of green space per se but the fact that children's access to such places is restricted by societal constraints (Gill, 2007; Malone, 2007; Freeman, 2010). It is not just physically green spaces that enhance happiness in the environment but sustainable, resilient and green cities generally. The study by Cloutier *et al.* (2014: 633) on happiness in US cities found 'positive associations between sustainable development and happiness on all scales'. They looked at a range of studies and their findings showed a positive relationship between factors such as liveability, walkability, urban design, greenness and happiness. They postulate that such cities and design features may encourage citizens to be more invested in their communities, which in turn contributes to positive social environments (Cloutier *et al.*, 2014: 643).

Convivial cities

Given the evidence that happiness for all citizens, especially children, is positively related to social connectivity and cohesion, it follows that those design features that encourage conviviality will also add to collective happiness. Such characteristics include:

- People gathering places: the city squares and market places
- Places that encourage pausing and interaction: provision of seats, cafes, stalls
- Places that attract: fountains, public art, street theatre

- Adaptable places: used for celebrations, games, quiet contemplation
- Places for all: children, elderly, disabled, dog walkers
- Well-designed places: attractive, comfortable, safe, good microclimate, acoustically pleasant, natural elements
- Well located places: central, easily accessible, good views
- Places at the right scale: not too large or small, human scale (after Shaftoe, 2008).

The creation of social opportunities is not confined to gathering spaces but can be enlarged to include the diverse range of spaces present in children's lives. Table 6.1 suggests some ways these spaces can be designed or have their environments enhanced to encourage happiness in children's lives. Although the spaces in the table do not on their own guarantee happiness, they can contribute to creating more convivial environments where children feel valued and that nurture social connections. Montgomery writes that 'we build the happy city by pursuing it in our own lives and, in so doing, pushing the city to change with us. We build it by living it' (2013: 321). Cities that encourage life, encourage happiness. The final section of this chapter considers how one city encourages happiness in its public life.

Fun in the city

In the research that exists on happiness and health, or more commonly on ill health, there is virtually nothing written about fun and how cities can foster 'fun'. Although we recognise 'fun' in real contexts, it can be hard to define. Carroll defined fun in the context of computer learning as follows: 'Things are fun when they attract, capture, and hold our attention by provoking new or unusual perceptions, arousing emotions . . . when they surprise us . . . when they present challenges or puzzles' (Carroll, 2004: 39). To Carroll's definition I would add that fun is active, rarely solitary and often infectious. Fun matters because fun is happiness at its most evident. Cities around the world are increasingly waking up to the need to encourage fun and to provide convivial events for their citizens. Some of these are traditionally organised events as in carnivals, pageants, fetes, parades and annual celebrations such as Christmas and New Year events; others are more organic and unpredictable like a snowfall that closes schools or a sporting event such as the Tour de France passing through the town.

The following examples come from Dunedin, the second-largest city (120,249 people in 2013; Statistics New Zealand, 2013) in New Zealand's South Island. Like most cities, Dunedin has a schedule of civically-supported events. These can be costly and are often the focus of budget debates, particularly when budget cuts are called for. However, given the benefits they bring to so many people, I would argue that their retention is essential even in times of financial hardship. Public events are essential to the wellbeing, health and happiness of the people, especially for children from low income families and are a key element of childhood memories. Such events must be free and, so far, all those profiled here in Dunedin have stood firm whilst other areas and events have seen public funding cuts.

Table 6.1 Creating better, happier environments for children

Type of space	Positive attributes
Homes	Homes that connect to each other, face each other, have communal spaces, low fences or walls that children can see over or, better still, have no physical boundaries on the front.
	Homes that face each other and relate well to communal spaces, such as green spaces or the street.
	Homes that aren't too big so children can easily see each other.
Streets	Streets that children can traverse safely and independently, that provide spaces for people to meet and chat, and where children get to know their community.
	Safe streets with slow, limited through-traffic, regular safe crossings, large pavements.
	Low traffic streets encourage interaction.
Transport	Regular public transport serving all neighbourhoods.
	Transport that encourages adults to use it rather than drive.
	Integrated systems to allow children to scoot, bike, walk between transport modes.
	Places to leave bikes and scooters safely.
People-safe environments	Communities that are safe for children to get around and encourage children's independence.
	There are adults around to provide support for children.
	Places where people know each other.
	Places where it is expected that children will be out and about.
Local services	Shops, doctors, swimming pools, schools and markets that cluster in the neighbourhood encouraging and maximising local patronage and that children can easily and independently access.
	Corner shops are especially important.
Play spaces	Playgrounds – formal and informal – are essential for children's wellbeing, developing physical skills, social interaction skills and just having fun.
Schools	Local schools attended by children from the same neighbourhood where they can travel to school together, have friends and meet up outside school.
	Where parents can gather and get to know each other.
	Schools that encourage use by children outside school hours and form neighbourhood hubs.
Convivial spaces	Places for special outings that reach out to children to stop, play, eat and generally have fun.
	Places designed to accommodate lots of people and different activities.
	Places with interesting and exciting happenings, and things children can watch, learn about and enjoy public life.
Green environments	Green spaces, healthy clean environments, natural features, space for children to hang out and space to play and have adventures.
	Trees and plants to beautify the neighbourhood and to provide shade.

All the events discussed here are free and open to all children. In addition to these council events are opportunities organised by groups, which often work at neighbourhood level such as school fairs, sports occasions or Christmas celebrations. Table 6.2 describes briefly some of the council's more popular city-wide events.

Table 6.2 Annual fun events organised or supported by Dunedin City Council

Organised event	Description
Jaffa Race	Sponsored by the local Cadbury Chocolate Factory, thousands of small, round orange chocolates and slightly larger purple chocolates are released at the top of the world's steepest street, watched by crowds to see which one will win the race to the bottom. Some of the larger ones bounce really high, few make it to the end and many children help themselves to the chocolates that fail to reach the end.
Mid-Winter Carnival	This event celebrates Matariki (Māori New Year festival) and the Winter Solstice. Held in mid-June, children and families parade in the dark around the central city square (actually an Octagon) carrying lanterns they have made in the preceding weeks. Other children and families line the Octagon to watch. Fireworks end the celebration.
Santa Parade	A well-loved annual Christmas event where local organisations prepare over 70 floats/trucks that parade through the main street accompanied by bands, streetwalkers, water squirting seals, dancers and Santa himself, with free chocolates thrown to the children in the waiting crowds.
Trolley Derby	A smaller event where children and some not-so-young people race homemade go-carts down a hilly street watched by crowds of enthusiasts.
Thieves Alley Market	Held on one Saturday each summer, the octagon and side streets are closed and filled with market stalls selling everything from local handicrafts to Peruvian handbags and sweets. Local organisations have stalls, there are rides, bouncy castles and free entertainment, including magic shows.
Chinese New Year	Celebrated with dragons, lions, fortune cookies, drums, a parade, free entry into the Chinese gardens and ending in fireworks. This celebration is an important family occasion for the Chinese community but celebrated by the wider community too.
Special Rigs	Now going for over 20 years some 150+ trucks all beautifully cleaned and polished circumnavigate the city beeping their horns whilst giving rides to 'special' children (children with disabilities, illnesses or experiencing hardship) and their carers/families. Children line the streets to watch these trucks, which include anything from fire engines to logging trucks, stock trucks and house removal vans.
Art Gallery	The city art gallery in recent years has hosted several innovative installations increasingly geared towards children, including Seung Yul Oh's interactive play exhibition (a huge climbing beanbag, a forest of yellow blow-up pillars and rocking, rotatable 'egg' birds); Inez Crawford's bouncy marae (like a bouncy castle but in the form of a marae, which is Māori meeting house); and a Lego city building exhibition involving Danish artist Olafur Eliasson providing thousands of white Lego bricks that visitors transform into an art city landscape. All of these were free and exceptionally popular attractions for the city's children, bringing laughter, fun and excitement into the gallery.

Table 6.2 Continued.

Organised event	Description
Polar Plunge	Now in its 86th year and run by the local Life Saving Club, children and adults brave the city's freezing seas to take a mid-winter plunge. Many dress in costume for the occasion and in recent years have been led into the sea by the Mayor.
Snow Days	Eagerly anticipated by the children because snow days usually mean a day off school and rare opportunities for sledging, building snow men and walks on snow-covered streets usually impassable to vehicles.
King Tide	These are very high tides that occur periodically during the year causing waves to crash against the sea wall sending sea spray cascading over the wall onto the walkways – to the great amusement of children and some adults, who can all end up quite soaked.

Conclusion: can cities be happy places?

The best cities are those whose citizens feel fulfilled and happy – they are not the richest, the largest or even the greenest cities but they are always cities that meet people's basic needs and that value all citizens. There have been a number of studies seeking to rank cities with reference to happiness (Leyden *et al.*, 2011) and a proliferation of rankings of cities based on indices of wellbeing, quality of life, liveability and some happiness quotient. Other studies attempt to identify the particular components that produce happiness in particular cities (Lima, 2014). Lima's study of happiness is situated within the context of attempts to define a 'Well-being Brazil' indicator in relation to civic spaces and urban policy. However, such attempts invariably suffer from problems of generalisation. Happiness is constructed in addition to physical factors from a range of personal and societal factors. At a general level, Rio de Janeiro may score well on a happiness scale in certain places and amongst some population groups, but for many residents poverty, isolation, factors of ethnicity and temperament may dominate their emotions and experiences. Although recognising elements of place specificity and personal experience are important in determining happiness, there are nonetheless some general negative and positive lessons that can still be drawn. If a city's children are cold, hungry, suffer from preventable diseases, are bored, frustrated, attend under-resourced schools and are precluded from independent travel by dangerous streets, the city and its citizens are unlikely to thrive. Happy children can be seen as the barometer of a good city and good health is an essential prerequisite for happiness.

Children seem to be predisposed towards a happy state. It can be argued that it takes a concerted effort of negativity to create unhappy children and research also seems to support the notion that happiness is an achievable state for most. Holder and Coleman (2009) reports that studies that use self-report and those that rely on other reports of children's happiness consistently state that children experience high levels of happiness. His own studies found over 90 per cent of children interviewed could be classed as happy. To see if happiness was just a condition of the wealthier Canadian cities, where the research was initially conducted, a further study was undertaken in North India. This study found 94 per cent of the children rated themselves in the top three happiest categories of the Faces Scale (a visual scale used to establish levels of happiness) and none of the children rated themselves in the three lowest categories of happiness (Holder *et al.*, 2012: 68). Given this positive propensity towards happiness it should not be too difficult for planners in less challenging environments to support happiness. Montgomery identified what he calls a 'recipe for urban happiness' for the city as follows:

 i   It should strive to maximise joy and minimise hardship
 ii It should lead towards health rather than sickness
 iii It should offer real freedom to live, move, build our lives as we wish
 iv It should build resilience against economic and environmental shocks
 v   It should be fair in the way it apportions space, services, mobility, joys, hardships and costs
 vi Most of all it should enable us to build and strengthen social bonds

Adapted from Montgomery, 2013: 43

Regeneration initiatives in many Western cities are directed at being 'happiness creators' as they move towards remaking more liveable neighbourhoods. These are neighbourhoods that maximise people contact through reduced traffic, have an emphasis on walkability, support mixed use and encourage use of local facilities, and provide resources to enhance the quality of homes, local services and the physical environment through greening and other initiatives. Unfortunately, in many urban environments past planning practices have acted against good 'happy' planning through prioritising traffic over people, removing or centralising services and facilities such as shops, swimming pools and even playgrounds and schools, and through allowing the construction of buildings and streets on an inhuman scale. Furthermore, regeneration initiatives can also in themselves have adverse effects where, for example, resident populations become displaced from the regeneration area due to rising levels of unaffordability.

The fundamental requisites for happy children in cities are relatively simple: good homes and services, good spaces that support play and independence, access to other children and people, and safe environments. Unfortunately, these are invariably areas in which governments seem to be becoming less willing to invest, especially for the more deprived and marginalised population groups, which includes children (Office of the Children's Commissioner, 2014; New Zealand Government, 2012). The basics must be present for all children for happiness to prevail and if politicians, planners, urban designers, architects and child advocates are to create cities for happy, healthy children they could do no better than to follow Bhutan's lead and ask of every development and every policy initiative 'will it increase people's happiness'?

References

Anakwenze, U. and Zuberi, D. 2013. Mental health and poverty in the inner city. *Health and Social Work*, 38(3), pp. 147–57.
Batcho, K.I., Nave, A.M., and DaRin, M.L. 2011. A retrospective survey of childhood experiences. *Journal of Happiness Studies*, 12, pp. 531–45.
Blazek, M. and Windram-Geddes, M. 2013. Thinking and doing children's emotional geographies. *Emotion, Space and Society*, 9, pp. 1–3.
Brockmann, H., Delhey, J., Welzel, C. and Yuan, H. 2009. The China puzzle: falling happiness in a rising economy. *Journal of Happiness Studies*, 10, pp. 387–405.
Carroll, J.M. 2004. Beyond fun. *Interactions*, 11(5), pp. 38–40.
Centre for Bhutan Studies & GNH Research. 2015. Home. Available at www.bhutanstudies. org.bt/home-2/ (accessed 29 August 2015).
Chaplin, L.N. 2009. Please may I have a bike? Better yet, may I have a hug? An examination of children's and adolescents' happiness. *Journal of Happiness Studies*, 10, pp. 541–62.
Cloutier, S., Larson, L. and Jambeck, J. 2014. Are sustainable cities 'happy' cities? Associations between sustainable development and human well-being in urban areas of the United States. *Environment, Development and Sustainability*, 16, pp. 633–47.
Conradson, D. 2012. Wellbeing: reflections on Geographical engagements. In: S. Atkinson, S. Fuller and J. Painter, eds. *Wellbeing and place*. Farnham: Ashgate, pp. 15–34.
Csikszentmihalyi, M. and Hunter, J., 2003. Happiness in everyday life: the uses of experience sampling. *Journal of Happiness Studies*, 4, pp. 185–99.
Demir, M. and Weitekamp, L.A. 2007. I am so happy 'cause today I found my friend: friendship and personality as predictors of happiness. *Journal of Happiness Studies*, 8(2), pp. 181–211.
Eloff, I. 2008. In pursuit of happiness: how some young South African children construct happiness. *Journal of Psychology in Africa*, 18(1), pp. 81–7.
England Marketing. 2009. *Report to Natural England on childhood and nature: a survey on changing relationships with nature across generations*. Available at http://publications. naturalengland.org.uk/publication/5853658314964992 (accessed 26 February 2017).
Fleuret, S. and Atkinson, S. 2007. Wellbeing, health, and geography: a critical review and research agenda. *New Zealand Geographer*, 63, pp. 106–18.
Freeman, C. 2010. Children's neighbourhoods, social centres to 'terra incognita'. *Children's Geographies*, 8(2), pp. 157–76.
Freeman, C. and Tranter, P. 2011. *Children and their urban environments: changing worlds*. London: Earthscan.
Freeman, C., Lingham, G. and Burnett, G. 2015. Children's changing urban lives: a comparative New Zealand–Pacific perspective. *Journal of Urban Design*, 20(4), pp. 507–25.
Freeman, C. van Heezik, Y. Stein, A. and Hand, K. 2016. Technological inroads into understanding city children's natural life-worlds. *Children's Geographies*, 14(2), pp. 158–74.
Gandelman, N., Piani, G., and Ferre, Z. 2012. Neighborhood determinants of quality of life. *Journal of Happiness Studies*, 13, pp. 547–63.
Garhammer, M. 2002. Pace of life and enjoyment of life. *Journal of Happiness Studies*, 3, pp. 217–56.
Gill, T. 2007. *No fear: growing up in a risk averse society*. London: Calouste Gulbenkian Foundation.
Graham, C., Higuera, L. and Lora, E. 2011. Which health conditions cause the most unhappiness? *Health Economics*, 20, pp. 1431–47.

Hellevik, O. 2003. Economy, values and happiness in Norway. *Journal of Happiness Studies*, 4, pp. 243–83.

Holbrook, A. 2008. *The green we need: an investigation of the benefits of green life and green spaces for urban dwellers' physical, mental and social health.* Epping, NSW: Nursery and Garden Industry.

Holder, M.D. 2012. Happiness in children. *Springer Briefs in Well-Being and Quality of Life Research.*

Holder, M.D. and Coleman, B. 2009. The contribution of social relationships to children's happiness. *Journal of Happiness Studies*, 10, pp. 329–49.

Holder, M.D., Coleman, B. and Singh, K. 2012. Temperament and happiness in children in India. *Journal of Happiness Studies*, 13, pp. 261–74.

Howell, A.J., Dopko, R.L., Passmore, H. and Buro, K. 2011. Nature connectedness: associations with well-being and mindfulness. *Personality and Individual Differences*, 51, pp. 166–71.

Kahn, P.H. and Kellert, S.R. eds. 2002. *Children and nature: psychological, sociocultural and evolutionary investigations.* Cambridge, MA: MIT Press.

Kearns, R. and Collins, D. 2009. Health geography. In: T. Brown, S. Mclafferty, and G. Moon, eds. *A companion to health and medical geography.* Hoboken, NJ: Wiley, pp. 15–32.

Khan, N.Y., Ghafoor, N., Iftikhar, R. and Malik, M. 2011. Urban annoyances and mental health in the city of Lahore, Pakistan. *Journal of Urban Affairs*, 34(3), pp. 297–315.

Korpela, K.M. and Yle'n, M. 2007. Perceived health is associated with visiting natural favourite places in the vicinity. *Health & Place*, 13, pp. 138–51.

Lachowycza, K. and Jones, A.P. 2013. Towards a better understanding of the relationship between greenspace and health: development of a theoretical framework. *Landscape and Urban Planning*, 118, pp. 62–9.

Lelkes, O. 2013. Minimising misery instead of maximising happiness? A new strategy for public policies. *Social Indicators Research*, 114, pp. 121–37.

Leyden, K.M., Goldberg, A. and Michelbach, P. 2011. Understanding the pursuit of happiness in ten major cities. *Urban Affairs Review*, 47(6), pp. 861–88.

Lima, I. 2014. Towards a civic city: from territorial justice to urban happiness in Rio de Janeiro. *European Journal of Geography*, 5(2), pp. 77–90.

Louv, R. 2005. *Last child in the woods: saving our children from nature-deficit disorder.* Chapel Hill, NC: Algonquin.

MacKerron, G. and Mourato, S. 2013. Happiness is greater in natural environments. *Global Environmental Change*, 23, pp. 992–1000.

Malone, K. 2007. The bubble-wrap generation: children growing up in walled gardens. *Environmental Education Research*, 13(4), pp. 513–27.

Mayer, F.S. and McPherson, F.C. 2004. The connectedness to nature scale: a measure of individuals' feeling in community with nature. *Journal of Environmental Psychology*, 24, pp. 503–15.

McCurdy, L.E., Winterbottom, K.E., Suril S. Mehta, S.S. and Roberts, J.R. 2010. Using nature and outdoor activity to improve children's health. *Current Problems in Pediatric Adolescent Health Care*, 40(5), pp. 102–17.

Mitchell, R. and Popham, F. 2008. Effect of exposure to natural environment on health inequalities: an observational population study. *Lancet*, 372(9650), pp. 1655–60.

Montgomery, C. 2013. *Happy city: transforming our lives through urban design.* New York: Farrar Strauss and Giroux.

New Zealand Government. 2012. *White Paper for Vulnerable Children*. Available at www. childrensactionplan.govt.nz/assets/Uploads/white-paper-for-vulnerable-children-volume-3.pdf (accessed 26 February 2017).

Nisbet, E.K., Zelenski, J.M. and Murphy, S.A. 2011. Happiness is in our nature: exploring nature relatedness as a contributor to subjective well-being. *Journal of Happiness Studies*, 12, pp. 303–22.

O'Campo, P., Salmon, C. and Burke, J. 2009. Neighbourhoods and mental well-being: what are the pathways? *Health & Place*, 15(1), pp. 56–68.

Office of the Children's Commissioner, The University of Otago's NZ Child and Youth Epidemiology Service, and the JR McKenzie Trust NZ. 2014. *Child Poverty Monitor, 2014 Technical Report*. Available at www.childpoverty.co.nz (accessed 26 February 2017).

Pearson, A.L., Ivory, V., Breetzke, G. and Lovasi, G.S. 2014. Are feelings of peace or depression the drivers of the relationship between neighbourhood social fragmentation and mental health in Aotearoa/NewZealand? *Health & Place*, 26, pp. 1–6.

Proctor, C.L., Linley, A. and Maltby, J. 2009. Youth life satisfaction: a review of the literature. *Journal of Happiness Studies*, 10, pp. 583–630.

Ryan, R.M., Weinstein, N., Bernstein, J., Brown, K.W., Mistretta, L. and Gagné, M. 2010. Vitalizing effects of being outdoors and in nature. *Journal of Environmental Psychology*, 30, pp. 159–68.

Sargeant, J. 2010. The altruism of pre-adolescent children's perspectives on 'worry' and 'happiness' in Australia and England. *Childhood*, 17(3), pp. 411–25.

Schaefer-McDaniel, N. 2009. Neighborhood stressors, perceived neighborhood quality, and child mental health in New York City. *Health & Place*, 15(1), pp. 148–55.

Shaftoes, H. 2008. *Convivial urban spaces: creating effective public spaces*. London: Earthscan.

Statistics New Zealand. 2013. *2013 Census QuickStats about a place: Dunedin City*. New Zealand Government. Available at www.stats.govt.nz/Census/2013-census/profile-and-summary-reports/quickstats-about-a-place.aspx?request_value=15022&tabname= (accessed 27 February 2017).

Tafarodi, R.M., Bonn, G., Liang, H., Takai, J., Moriizumi, S., Belhekar, V. and Padhye, A. 2012. What makes for a good life? A four-nation study. *Journal of Happiness Studies*, 13, pp. 783–800.

Uchida, Y., Norasakkunkit, V. and Shinbo, K. 2004. Cultural constructions of happiness: theory and empirical evidence. *Journal of Happiness Studies*, 5, pp. 223–39.

United Nations. 2014. *World's population increasingly urban with more than half living in urban areas*. Available at www.un.org/en/development/desa/news/population/world-urbanization-prospects-2014.html (accessed 26 February 2017).

United Nations. n.d. *Convention on the Rights of the Child – Adopted and opened for signature, ratification and accession by General Assembly resolution 44/25 of 20 November 1989*. New York: United Nations.

Uusitalo-Malmivaara, L. and Lehto, J.E. 2013. Social factors explaining children's subjective happiness and depressive symptoms. *Social indicators Research*, 111, pp. 603–15.

Part II

Home and away

7 Mobilising children

The role of mobile communications in child mobility

Bjorn Nansen, Penelope Carroll, Lisa Gibbs,
Colin MacDougall and Frank Vetere

The role of mobile phones in children's mobility remains underexplored despite their growing take-up and use. The few extant studies of the phenomenon of children's mobile communication are dominated by analyses of parent–child relations and findings that mobile phones extend both child travel and parental intrusion (e.g. Fotel and Thomsen, 2004; Williams and Williams, 2005; Malone, 2007). There is, therefore, a clear need to consider the wider contexts, meanings and practices associated with children's increasing and everyday uses of mobile technologies. In this chapter, we explore ways that mobile phones mediate children's everyday mobility, extending research in the geographies of health literature that has addressed questions of mobile phone use in the context of children's independent mobility. We report on two qualitative studies examining how children and parents perceive and use mobile phones to negotiate mobility: the first in an inner urban area of Moreland in Melbourne, Australia, during 2011 and 2012; and the second in inner city Auckland, New Zealand, in 2012. A total of 48 Australian children and 40 New Zealand children aged 9–12 years – a transitional age in social and educational terms – along with parents, were included in the research.

In the chapter, we use the term 'companion device' to refer to the growing significance of mobile phones in accompanying children when they are out and about, and introduce the idea of media ecologies as a helpful theoretical approach. This allows us to situate mobile phone use within a broader field of social and technological relations characterising children's mobility negotiations and communications. Thus, we consider how children's everyday mobility is negotiated through mobile phones within the contexts of family rules and routines, cultural influences, peer social connections and neighbourhood environments. We detail three themes of mobile device ecologies – adoption, affordance and appropriation – in order to analyse the shifting significance of these companion devices in children's everyday mobility.

Background and theoretical approach

The implications of mobile phone usage have been explored within the contexts of family communication, interaction and organisation (e.g. Green, 2002, 2003; Licoppe, 2004; Christensen, 2009; Davis *et al.*, 2011), yet the role of mobile technologies in child mobility remains underexplored within health geography literature. Here the focus tends to have been on the broader social, physical and policy environments that have contributed to a dramatic decline in children's everyday mobility, and particularly independent mobility (e.g. O'Brien *et al.*, 2000; Malone, 2007; McDonald, 2007; Thomson, 2009; Zubrick *et al.*, 2010; Carroll *et al.*, 2015).

Independent mobility – defined as freedom to travel or move about neighbourhoods without adult supervision (Tranter and Whitelegg, 1994; Shaw *et al.*, 2013) – is considered important for physical health, helping children to incorporate active transportation such as walking and cycling into their daily travel routines (e.g. Carver *et al.*, 2008; Garrard, 2009; Thomson, 2009). It is also valued for fostering children's wellbeing more broadly, increasing their spatial, personal and social skills, and providing children with opportunities to access public space for play, recreation and citizenship (e.g. Malone, 2007; Ross, 2007; Skelton, 2009; Zubrick *et al.*, 2010). Studies have shown that factors contributing to a reduction in children's independent mobility include changes to built environments and transportation policies (e.g. Carver *et al.*, 2008; Whitzman *et al.*, 2009); shifts in parental perceptions and rules (e.g. Valentine and McKendrick, 1997; Prezza *et al.*, 2005; Zubrick *et al.*, 2010); and changing cultural norms, values and lifestyles (e.g. Gill, 2007; Malone, 2007; Thomson, 2009). Less is known, however, about the increasing role of mobile communication technologies in shaping children's contemporary independent mobility. In this chapter we add to the evidence by reporting on two qualitative studies exploring how children and parents perceive and use mobile phones to negotiate everyday mobility.

Studies of mobile phone use in the context of independent mobility have tended to focus on teenage practices (e.g. Green, 2003; Williams and Williams, 2005), and research on pre-teen children's mobile phone use is sparse. Although this may be due to the historical lack of mobile ownership or use by pre-teens, evidence shows an ever-increasing ownership of mobile devices by children – and by increasingly younger children (Australian Bureau of Statistics (ABS), 2012). Extant research has focused on mobile phone use in child–parent relationships rather than use as part of children's broader social relationships. The findings of much of this research note that children gain a degree of empowerment and autonomy from having a phone, and that the mobile phone increases their spatial mobility; but at the cost of parental surveillance that children view as an invasion of their space (e.g. Fotel and Thomsen, 2004; Williams and Williams, 2005; Malone, 2007). Phones may therefore extend boundaries, but they simultaneously stretch parental authority and intrusion. Issues of surveillance notwithstanding, recent research on children's mobility is beginning to consider the ways mobile devices are reconfiguring parent–child relationships, with the potential to help reverse trends of reduced child mobility (e.g. Kullman, 2010; Shaw *et al.*, 2013; Nansen *et al.*, 2014).

Our goal is to explore the role played by mobile phones in mediating children's travel, situating the use of mobile phones within the contexts of family rules and routines, cultural influences, peer social connections and neighbourhood environments. We build on recent cultural geography literature that recognises that children increasingly live in a multi-mediated world, where their everyday mobility is constructed and sustained through the quality of relations with, or attachments to, bodies, places and technologies (Ross, 2007; Kullman, 2010); where technologies quietly surround and attach themselves to their daily practices (e.g. Symes, 2007); and where the companionship of human others – family, siblings, friends and passers-by – as well as 'non-human' others, such as pets and technologies, help foster children's independent mobility (Mikkelsen and Christensen, 2009; Nansen *et al.*, 2014).

In building on this research in cultural geography, we consider the companionship role of mobile phones in mediating children's everyday mobility. The term companion device recently emerged in the lexicon of the consumer electronics industry to refer to a second screen or additional device (e.g. laptop, tablet, smart phone) that is kept close at hand and is used in conjunction with television viewing, therefore becoming embodied in accompanying and adding a layer of complexity to this activity. We appropriate the term to extend it beyond the home environment to describe children and young people's relations to, and through, mobile phones within their everyday mobility. We suggest that the term companion device speaks to qualities of use, experience and meaning attached to mobile phones by children.

Furthermore, we adopt an ecological framework to consider the broader social and technological relationships characterising children's mobility. A socio-ecological approach has been used in health geography to examine the multiple and interrelated factors influencing health, including individual and interpersonal relations; institutional and regulatory rules; physical settings and environments; social norms and cultural contexts (e.g. Kearns, 1993; Garrard, 2009). An ecological approach has also been used within media and communication studies to explore the interrelations between multiple technologies, users and contexts of use (e.g. Strate, 2004; Fuller, 2007; Hearn and Foth, 2007; Shepherd *et al.*, 2007). Here, the ecological metaphor shifts the focus away from studies of individual devices to encompass systems and contexts of media and communications interaction. We suggest an ecological framework is useful for explicitly drawing attention to the role of mobile phones in shaping children's mobility, without losing sight of the larger contexts in which they are situated.

Rather than treating mobile phones in isolation, the ecological approach suggests ways they mediate, amend and reconfigure practices of child mobility. We detail three themes of mobile device ecologies – adoption, affordance and appropriation – to analyse the shifting significance of these companion devices in children's everyday mobility communications and negotiations, that is, the ways in which these companion devices are adopted and appropriated through their affordances.

The psychologist James Gibson (1977, 1979) coined the term 'affordances' to describe physical properties embedded within an object, technology or environment that enabled an individual to perform some action. It has been expanded to include properties (physical and symbolic) that users themselves identify or perceive (e.g. Norman, 1988; Costall, 1995). Kullman (2010) argues that the affordances of travel technologies (such as mobile devices) serve a material and performative role as a resource for mediating children's mobility transitions and as an aid to actively assembling contemporary forms of child mobility. This relational and interdependent understanding of affordance is made clearer through the concept of appropriation, whereby technologies exceed their intended design through user appropriation and innovation (e.g. Marvin, 1988; Verbeek, 2005; Wajcman, 2008). Pain *et al.* (2005), for example, argue that instead of bringing fundamental changes to the lives of young people, mobile phones have been appropriated into the existing ties, struggles and surveillance between young people and adults. That is, although the contactability and connectedness afforded by mobiles may expand the boundaries and modes of parent–child interactions or mobility negotiations, they fail to actually transform the dynamics of child mobility.

In contrast, Williams and Williams (2005) argue that in facilitating parent–child negotiations that are no longer restricted to the home, mobile phones afford new forms of child agency that alter the experience and character of mobility. Mobile phones stretch these relations over distance and enable negotiations within a broader spatial framework in which children are increasingly playing an active role. Patterns of appropriation show that young people integrate and adapt mobile devices as part of a wider communications ecology, with the mobile phone used for brief communicative gestures (Licoppe, 2004). Child mobility research has also shown that children appropriate mobile devices to subvert their intended purposes of adoption, such as remote parenting or surveillance. Young people decide how much to share with their parents (Pain *et al.*, 2005) or switch off their phone to avoid contact with their parents (Williams and Williams, 2005).

A brief introduction to the research methods

Both studies used mixed methods child-centred approaches. In the Australian study, ethnographic observations preceded focus group discussions with children in groups of 5–8 (n = 48) centred around two visual photo-ordering exercises using images depicting places of travel and objects mediating their mobility (Morrow, 2001; Punch, 2002). A Year 6 to Year 7 cohort (10 and 11-year-olds) was selected because research shows the transition from primary to secondary school is a significant stage in children's mobility development (Shaw *et al.*, 2013) and an important period in the acquisition of mobile phones (ABS, 2012). Group discussions were followed by participant-guided mobile methods (Block *et al.*, 2014), which in this instance included 10 children from the focus groups individually taking researchers on daily travel journeys predominantly to and from school, but also to places such as shops and parks. Researchers used the same mode of transportation as participants (e.g. bus, walking, cycling). Parent and teacher interviews provided context and triangulation (for detailed methods description see Gibbs *et al.*, 2012; Nansen *et al.*, 2014).

The New Zealand study investigated how children and parents experience different neighbourhood environments and the focus was on relationships between urban design attributes, safety perceptions and children's independent mobility and physical activity levels. Here we present data related to mobile phone use collected from 34 parents and 40 children aged 9–12 years living in inner-city apartments and townhouses during home-based semi-structured interviews with parents/caregivers and children; and neighbourhood go-along walking interviews with children (for full details of methodology see Carroll *et al.*, 2015). As with the Australian study, the children were on the cusp of greater freedom as they transitioned from primary school (Years 5–6) to intermediate/junior secondary school (Years 7–8).

Children's adoption of mobile devices

Our research supports previous findings that children often received their first mobile phones in response to increased travel unaccompanied by parents, in particular to and from school (Haddon and Vincent, 2007). These studies found that this was during the transition between primary and secondary school for the Australian cohort, and between primary and intermediate/junior high school for New Zealand participants. Mobile phones were principally given to children as travel companions by parents who wanted them to be contactable (Ling, 2004; Haddon and Vincent 2007; Wajcman *et al.*, 2007) and initially the children's use of these was limited:

> Child, 10 years (Year 6, Australia): Mine's just for emergencies.

> Child, 10 years (Year 6, Australia): I don't call my friends just my parents.

> Father (of Year 6 child, Australia): We will get him a mobile phone next year to be able to stay in touch.

When speaking about younger children being given a mobile phone once they transitioned from primary school, the main reason parents gave was the greater distance that children would travel to school unsupervised by an adult, as exemplified in the following account:

> Liz, mother of Michael, 10 years (Year 6, NZ): It's a bit early for him to have a mobile phone. His school is close to home so he doesn't need it yet. . . but he might need one when he goes to intermediate.

Children appeared to accept this delay in acquiring a mobile phone and some younger children talked of getting one for intermediate (junior high) school. Some younger children who owned phones did not regularly carry them because they were not seen as a necessary technology for mediating shorter travel journeys:

> Jess, 10 years (Year 6, NZ): Oh no, not all the time.

> Luke, 11 years (Year 7, NZ): I take it sometimes.

Children knew the parental contact enabled by mobile phones meant adoption was often accompanied by forms of 'remote control' (Fotel and Thomsen 2004), and an 'invasion' of their space (Williams and Williams, 2005), particularly through having to communicate their arrival at school:

> Child, 10 years (Year 6, Australia): When we leave our house 'cause our Mum's gone to work we have to say we've left and it's like 8.15 or something and then when we get to school we just say 'we're at school'.

Nevertheless, these children also felt that carrying a mobile phone provided confidence and a reciprocal sense of safety or security through the ability to contact parents or other adults if they felt unsafe, and even for companionship:

> Child, 10 years (Year 6, Australia): I text to let my parents know where I am.

> Child, 10 years (Year 6, Australia): If you are in danger you can call someone; or just to give you some company.

Adoption of mobile phones by children shows that whilst parents often delayed their provision until children transitioned to intermediate/junior high or secondary school, at these times of increased independent mobility they were generally perceived as valuable devices and travel companions within a wider ecology of companions and infrastructure supporting child mobility.

Affordances of mobile devices

Ambivalence about parental contact or monitoring through mobile phones became more pronounced as children became older, had their devices longer and transitioned into secondary school. Some Year 7 children resented the rules for mobile phones and continued intrusion and remote control from parents:

> Child, 11 years (Year 7, Australia): I'm not allowed to leave the house without my phone, like 'just take your phone in case you get lost'. I'm not going to get lost!

> Child, 11 years (Year 7, Australia): My mum is so paranoid that I'm going to get in trouble she makes me call before I leave, when I get there, when I'm going to be back, what I'm having for lunch and dinner, if I'm staying there overnight!

Yet, older children also recognised that mobile phones enabled increased independent mobility. They acknowledged the reciprocal benefits of convenience in negotiating with parents using mobile phones – superseding the limits of located legacy landlines – which contrasts with the importance of safety identified by younger children:

> Child, 11 years (Year 7, Australia): With a mobile you can tell your parents where you are.

Child, 11 years (Year 7, Australia): If I'm at a friend's house I call them if I want to be picked up.

They still used mobile phones for parental contact and safety, but in addition they could reschedule or renegotiate meeting times/places with parents, thereby helping to mediate an ecology of mobility that enabled greater freedom:

Child, 11 years (Year 7, Australia): If your parents say to you be home at like 7 o'clock, you can call and tell them you'll be late.

Child, 11 years (Year 7, Australia): Instead of going straight home, you can say you are going out with your friends.

Safety, convenience and surveillance were all seen as affordances by parents who favoured their children carrying around mobile phones. For instance, if a child got hurt climbing trees or falling off scooters they could phone for help. Thus, they can be seen as both an enabler and a safety cushion:

Jane, mother of Nick, 11 years, (Year 6, NZ): Quite often Nick rings me on his cellphone and says his friend's fallen off his scooter and he's got scabby knees and we're on the corner of such and such and such and such and you have to run up with plasters and towels in the car.

Convenience involved parents being able to readily contact children – and especially when previously agreed arrangements changed:

Sue, mother of Sarah, 9 years (Year 5, NZ): You can communicate with them easily. . . but it's about convenience, not safety.

Lyn, mother of James, 11 years (Year 7, NZ): It's a convenience if plans change, if I don't know what is happening with my day.

Parental peace of mind was another reason parents gave for supplying their children with a mobile phone and having it accompany them in public:

Bev, mother of David, 11 years (Year 7, NZ): If he wants to go into town and things it worries me if he doesn't have a phone. I'm concerned 'cause I can't get hold of him.

Sally, mother of Sam, 9 years (Year 5, NZ): I feel more comfortable.

This meant they could be in contact with their children from a distance, know where they were and that they were fine:

Lea, mother of Sophie, 10 years (Year 6, NZ): She must always have her cellphone on her and text me when she's coming home. If she's coming home on the bus she has to text me when she gets on the bus so that I know that she's on the bus.

However, safety was also offered as a concern and reason by parents for children not to carry a mobile phone. In this case mobile phones were seen as an unsafe companion because their material value could put a child at risk:

> Judy, mother of Peter, 10 years (Year 6, NZ): I am scared they will be robbed.

Children were also aware that the affordances of connectivity offered by mobile phones could lead to distraction. Here the ease of using a mobile phone in situations whilst on the move was seen to pose potential traffic safety dangers:

> Child, 11 years (Year 7, Australia): Phones can be helpful but if you are using it wrong. . . Yeah like if you are walking across a road.

> Child, 11 years (Year 7, Australia): If you are texting while you are walking: boom!

Parents expressed similar ambivalence about mobile phone companionship, discussing how the range of functions afforded by mobile phones, such as texting or playing games, could distract children:

> Ange, mother of Jim, 10 years (Year 6, NZ): It would be more dangerous to have a cellphone because they'd be like texting and walking and scootering. . . and playing games and not actually looking where they were going.

Some parents spoke about the affective dimensions of mobile phone companionship, expressing concern that a mobile phone could provide a false feeling of security and that other measures such as learning how to deal with difficult situations and having adults around that they could turn to if they were concerned were important to ensure children were safe:

> Rose, mother of Mark, 10 years (Year 6, NZ): It's not guaranteed if they have a cellphone they will be safer. . .that's why I teach Jim to understand the situation, how to deal with it. . . I'll ask him, 'if you don't have a cellphone, what should you do'?

One parent spoke about how he did not think a mobile phone would save his daughter in a 'stranger-danger' situation:

> Simon, father of Rianna, 11 years (Year 6, NZ): If something was to happen to her, by the time she'd pulled it out, she would have been for it anyway. . .

Thus, both children and parents were ambivalent about the affordances offered by mobile phones as companion devices. They recognised the merits of children – especially younger children – having mobile phones, acknowledging the affordances they offered as companions to child travel, whilst questioning their utility for actually ensuring children's safety within the wider ecology of infra-structures supporting child mobility.

Appropriating mobile devices

Following the transition to secondary school, students viewed phones as essential for their daily lives and their mobile devices became entwined with peer relation-ships and ecologies of communication. Appropriations grew to include organising schedules with friends, the use of built-in cameras, music players, casual gaming and web browsers, as part of mobile companionship mediating daily travel:

> Child, 11 years (Year 7, Australia): . . .it's like essential.
>
> Child, 11 years (Year 7, Australia): We text each other to work out if we are going to meet on the second-last carriage or whatever, or if we are going to meet at McDonalds where the train stops.
>
> Child, 11 years (Year 7, Australia): And you can play games. . . Like Angry Birds.

In appropriating mobile devices within their everyday mobility, children devel-oped more nuanced mobile phone engagements and discussed having to negotiate material aspects of mobile phones, including their physical presence, battery life and affordability:

> Child, 11 years (Year 7, Australia): You'll have your mobile phone always on you in your bag.
>
> Child, 11 years (Year 7, Australia): My battery always dies.
>
> Child, 11 years (Year 7, Australia): Credit can be annoying. . . my friend keeps texting me but I don't have enough credit to text him back.

Children also appropriated the mobile phone within parental negotiations for increased independence, leveraging the perceived affordances noted previously. They viewed parental rules about safety as reasonable, but attempted to negotiate or challenge rules they saw as unnecessary or overly restrictive. Tactics included approaches such as making a claim for maturity, nagging or selecting which parent to approach, but also involved mediators of mobility that parents promoted for safety. Thus, chil-dren used mobile phone companionship as a tool to negotiate mobility with parents:

> Child, 11 years (Year 7, Australia): I have my phone and say, 'just give me a ring'.
>
> Child, 11 years (Year 7, Australia): I tell them that I'll call you when I get there.

Children spoke about their parents allowing them greater spatial mobility if they had a mobile phone with them. They gave instances of places they were allowed to go on foot, on scooters and on bicycles around the city without supervision as long as they agreed to keep in touch and be back home at an agreed time, and of negotiating spatial and temporal changes by calling or texting their parents from their phone:

> Child, 11 years (Year 7, Australia): If I want to go shopping I text her. Sometimes she says no, sometimes she says yes.
> Lucy, 11 years (Year 6, NZ): If we decided to go up P Road I would call her.

With many parents working, children were sometimes able to play out and about with friends as long as they kept parents informed of their whereabouts and were home by an agreed-upon time. Having a phone could thus also afford temporal flexibility:

> Matt, 10 years (Year 6, NZ): If mum wants us back, well she can ring us, 'cause we'll take our phones.

Nevertheless, several parents spoke about a continued wish to keep their children under surveillance when they were out and about:

> Kath, mother of Jon, 10 years (Year 6, NZ): When he comes out of school, if he hasn't called me by ten past three, I'm calling him. . .

Other parents expressed concern about the limits of their surveillance capabilities once mobile phone use becomes habituated and increasingly multi-functional within the communication ecologies of young people. This had unexpected financial implications for some parents:

> Gill, mother of Aaron, 9 years (Year 5, NZ): You can't control whether they will use them to play games or whatever, so I worry about that.
> Rachel, mother of Danny, 10 years (Year 6, NZ): He's shocking; I've had huge phone bills because he's downloaded stuff.

Thus, we see a progression in the sophistication and usage of mobile phones as children get older, and whilst mobile phones become indispensable as companion devices for younger people, their diverse modes of appropriations within their peer communications ecology opens up new terrains of concern for parents.

Conclusion

Both studies demonstrate the role of the mobile phone as an important mediator in parent–child negotiations relating to child mobility. The findings from these Australian and New Zealand studies were consistent in showing that mobile phones were a resource for supporting mobility, typically provided to children when they transitioned from primary school, as a means of contact with parents (see also Ling, 2004; Haddon and Vincent, 2007; Wajcman *et al.*, 2007).

They helped to assemble children's mobility through the cooperation and assistance of a range of people, objects and environments (Kullman, 2010; Nansen *et al.*, 2014). In both studies, children, peers and parents collaborated with and through companion devices to provide real and virtual visibility of their location and activities (Ross, 2007; Mikkelsen and Christensen, 2009; Kullman, 2010). Furthermore, mobile phones exceeded their initial function as travel companions to become integral technologies within the communications ecology that stretched the temporal and spatial boundaries of children and young people's social lives.

'Companion devices' such as mobile phones play an increasing role in accompanying children on their everyday travel. We found that mobile phones were often handed down from parents specifically in relation to children's developing mobility and they mediated the connected presence of others at a distance. Primary-school students viewed mobile phones as helpful for contacting their parents, thus offering security or safety (see also Christensen, 2009; Nansen *et al.*, 2014). Although both parents and children recognised the potential for phones to distract and thus potentially reduce children's safety as pedestrians, they described mobile phones as important mobile mediators enabling a sense of security, parental contact and child independence. Carrying a companion device to remain contactable provided a reciprocal sense of comfort and security in which children felt safer knowing they were able to access their parents, and thus parents were not simply remotely monitoring their children but able to help support their children's developing mobility through a connected presence (Christensen, 2009). The surveillance capacities of mobile phones can be viewed as part of a larger terrain of mutual surveillance, which in practice works as a collaborative rather than controlling arrangement (Green, 2002). The phone is part of a wider ecology of people, technologies and places that help to support children's mobility and potentially increased physical activity.

Older children described the more nuanced engagement they developed with and through their mobile phones as these companion devices became embedded in their everyday lives and communication ecologies over time (e.g. Symes, 2007). Furthermore, children began to appropriate mobile phones within a culture in which hanging out with and through these companion devices was both ordinary and sophisticated. The associated costs of increased phone usage and access to online games had unanticipated financial implications for some parents. Children used mobile phones beyond parental interactions within ecologies of communication to socialise, make arrangements with peers, access entertainment and navigate urban environments. Children also leveraged their competency with mobile phones in negotiations with parents to further their ambitions for increased travel in public space. Rather than a digital leash for parental monitoring of children, mobile devices emerged as part of a repertoire of resources for scaffolding children's mobility, supporting them as they ventured further afield.

Over time, the affordances of mobile phones facilitated contact and convenience, but also concern about potential unintended hazards threatening physical and personal safety. Parents in New Zealand anticipated some of the unintended consequences of mobile phones – specifically that phones could be a target for theft, they could give children a false sense of security and they may alter the nature of children's play and social activities. The fears of parents and the appropriation of affordances by children reinforce the relationship between children's agency, parental power and their role as protector. Children and parents alike could predict how mobile phones would contribute to a shift in power to children, which was reflected in the fears expressed by parents in New Zealand.

Taken together, the studies reflect the importance of research into the relationship between mobile phones and the independent mobility of children, particularly in the transition from primary to intermediate or secondary school. Early ideas of a parentally-controlled electronic leash have been replaced by the acknowledgement of affordances offered by mobile devices and their appropriation by children as travel companions. Some of these are expected consequences of the introduction of new technologies, such as increased active transport (Shaw *et al.*, 2013). Other unintended consequences are that while researchers and parents expected mobile phones to enhance children's safety, children themselves appropriated the devices over time to increase their agency in negotiations with their parents (Valentine, 1997; Pain *et al.*, 2005), to renegotiate mobility and independence in response to opportunities that emerged throughout the day (Williams and Williams 2005) and to support their own interests in relation to entertainment and communicating with friends (Haddon and Vincent 2007; Symes, 2007).

These contradictions highlight how mobile devices do not serve a singular or straightforward function. Although mobile phones may be contributing to shifts in the patterns and landscapes of negotiation between parents and children over mobility (Shaw *et al.*, 2013), they need to be considered within the wider social and communicative ecologies of use. Research examining children's everyday uses of mobile phones requires more integrated and contextual approaches that place the use of mobile phones in the broader dynamics of daily life and communication (e.g. Haddon and Vincent, 2007; Kullman, 2010): why they are adopted, what do they afford and how they are appropriated? We propose that an understanding of wellbeing shaped through both social and technical determinants of an ecology approach may inform a cultural shift in the historical decline of children's movement in public space.

In conclusion, this research supports the evidence that mobile phones have become central to parent–child negotiations relating to children's independent mobility, and thus their health and wellbeing. Not only do they afford a sense of comfort and security to both parents and children, they also provide a means of increasing children's independence over time. This is especially the case as children negotiate new arrangements with parents in response to emerging opportunities while out and about. This research also points to the growing significance of mobile phones in accompanying children when they are out and about. It has demonstrated the ways their adoption, affordances and appropriation exceed strictly parent–child relations, becoming entangled in wider relational ecologies characterising children's mobility negotiations and communications.

Given the recognition of the importance of independent mobility for children's wellbeing and healthy development – and the role of mobile phones in fostering this – these companion devices can arguably be considered health promoting, at least within the context of children's everyday mobility practices. The sense of security afforded to both parents and children appears to allow greater independent mobility, which is in turn acknowledged as important for increasing physical activity levels and, more broadly, developing children's spatial and social skills. Thus, the negative aspects of children's mobile phone usage voiced by parents (e.g. children appropriating their mobile phones for their own purposes, having a false sense of security and potentially being dangerously distracted while out and about) appear to be outweighed by health and wellbeing benefits. Rather, the adoption, affordances and appropriation of mobile phones by children help to scaffold their opportunities for negotiated independent mobility and engagement with peers and local environments.

Acknowledgements

The Australian component of this work was supported by the Victorian Health Promotion Foundation (VicHealth), under their Innovation grant scheme, and an Australian Research Council (ARC) Discovery Early Career Researcher Award (DE130100735). School participation was supported through the Department of Education and Early Childhood Development. Authors Gibbs and MacDougall also gratefully acknowledge support from the Jack Brockhoff Foundation. The New Zealand component was funded through the Marsden Fund (MAU1011). The authors would like to express their appreciation to the children, families, teachers and schools involved in these studies for their generous contribution of time and experiences.

References

Australian Bureau of Statistics (ABS). 2012. *Children's participation in cultural and leisure activities*. Australia: ABS.

Block, K., Gibbs, L., Snowdon, E. and MacDougall C. 2014. Participant guided mobile methods: investigating personal experiences of communities. *Sage Research Methods Cases*. London: SAGE.

Carroll, P., Witten, K., Kearns, R. and Donovan, P. 2015. Kids in the city: children's use and experiences of urban neighbourhoods in Auckland, New Zealand. *Journal of Urban Design*, 20(4), pp. 417–36.

Carver, A., Timperio, A. and Crawford, D. 2008. Neighbourhood road environments and physical activity among youth: The CLAN study. *Journal of Urban Health*, 85, pp. 532–44.

Christensen, T. P. 2009. 'Connected presence' in distributed family life. *New Media Society*, 11(3), pp. 433–51.

Costall, A. 1995. Socialising affordances. *Theory and Psychology*, 5, pp. 467–81.

Davis, H., Francis, P., Nansen, B. and Vetere, F. 2011. Family worlds: technological engagement for families negotiating urban 'traffic'. In M. Foth, L. Forlano, C. Satchell, and M. Gibbs, eds. *From social butterfly to engaged citizen*. Cambridge, MA: MIT Press, pp. 217–34.

Fotel, T. and Thomsen, U. 2004. The surveillance of children's mobility. *Surveillance & Society*, 1(4), pp. 535–54.

Fuller, M. 2007. *Media ecologies: materialist energies in art and technoculture*. Cambridge, MA: MIT Press.

Garrard, J. 2009. *Active transport: children and young people: an overview of recent evidence*. Melbourne: Victorian Health Promotion Foundation.

Gibbs, L., MacDougall, C., Nansen, B., Vetere, F., Ross, N., Danic, I., McKendrick, J. and LaMontagne, T. 2012. *Stepping out: children negotiating independent travel*. VicHealth report. Jack Brockhoff Child Health and Wellbeing Program, University of Melbourne.

Gibson, J. 1977. The theory of affordances. In R. Shaw, and J. Bransford, eds. *Perceiving, acting, and knowing: toward an ecological psychology*. Hillsdale, NJ: Lawrence Erlbaum, pp. 67–82.

Gibson, J. 1979. *The ecological approach to human perception*. Boston, MA: Houghton Mifflin.

Gill, T. 2007. *No fear: growing up in a risk-averse society*. London: Calouste Gulbenkian Foundation.

Green, N. 2002. Who's watching whom? Monitoring and accountability in mobile relations. In B. Brown, N. Green and R. Harper, eds. *Wireless world: social, cultural and interactional issues in mobile technologies*. London: Springer, pp. 32–45.

Green, N. 2003. Outwardly mobile: young people and mobile technologies. In J. Katz, ed. *Machines that become us: the social context of personal communication technology*. Piscataway, NJ: Transaction, pp. 201–18.

Haddon, L. and Vincent, J. 2007. *Growing up with a mobile phone – learning from the experiences of some children in the UK. A report for Vodafone*. Digital World Research Centre.

Hearn, G. N. and Foth, M. 2007. Communicative ecologies: editorial preface. *Electronic Journal of Communication*, 17(1–2).

Kearns, R. A. 1993. Place and health: towards a reformed medical geography. *The Professional Geographer*, 45(1), pp. 139–47.

Kullman, K. 2010. Transitional geographies: making mobile children. *Social & Cultural Geography*, 11(8), pp. 829–46.

Licoppe, C. 2004. Connected presence: the emergence of a new repertoire for managing social relationships in a changing communication technoscape. *Environment and Planning D: Society and Space*, 22, pp. 135–56.

Ling, R. 2004. *The mobile connection. the cell phone's impact on society*. San Francisco, CA: Morgan Kaufmann.

Malone, K. 2007. The bubble-wrap generation: children growing up in walled gardens. *Environmental Education Research*, 13(4), pp. 513–27.

Marvin, C. 1988. *When old technologies were new*. New York: Oxford University Press.

McDonald, N. 2007. Active transportation to school: trends among US schoolchildren, 1969–2001. *American of Preventative Medicine*, 32, pp. 509–16.

Mikkelsen, M. R. and Christensen, P. 2009. Is children's mobility really independent? A study of children's mobility combining ethnography and GPS/mobile phone technologies. *Mobilities*, 49(1), pp. 37–58.

Morrow, V. 2001. Using qualitative methods to elicit young people's perspectives on their environments: some ideas for community health initiatives. *Health Education Research, Theory and Practice*, 16(3), pp. 255–68.

Nansen, B., Gibbs, L., MacDougall, C., Vetere, F., Ross, N. and McKendrick, J. 2014. Children's interdependent mobility: compositions, collaborations and compromises. *Journal of Children's Geographies*, 13(4), pp. 467–81.

Norman, D. 1988. *The design of everyday things*. New York: Doubleday.

O'Brien, M., Jones, D., Sloan, D. and Rustin, M. 2000. Children's independent spatial mobility in the urban public realm. *Childhood*, 7(3), pp. 257–77.

Pain, R., Grundy, S., Gill, S., Towner, E., Sparks, G. and Hughes, K. 2005. 'So long as I take my mobile': mobile phones, urban life and geographies of young people's safety. *International Journal of Urban and Regional Studies*, 29, pp. 814–30.

Prezza, M., Alparone, F., Cristallo, C. and Luigi, S. 2005. Parental perception of social risk and of positive potentiality of outdoor autonomy for children: the development of two instruments. *Journal of Environmental Psychology*, 25, pp. 437–53.

Punch, S. 2002. Youth transitions and interdependent adult–child relation in rural Bolivia. *Journal of Rural Studies*, 18, pp. 123–33.

Ross, N. J. 2007. 'My journey to school. . .': foregrounding the meaning of school journeys and children's engagements and interactions in their everyday localities. *Children's Geographies*, 5(4), pp. 373–91.

Shaw, B., Watson, B., Frauendienst, B., Redecker, A., Jones, T. and Hillman, M. 2013. *Children's independent mobility: a comparative study in England and Germany (1971–2010)*. London: Policy Studies Institute.

Shepherd, C., Arnold, M., Bellamy, C. and Gibbs, M. 2007. The material ecologies of domestic ICTs. *The Electronic Journal of Communication*, 17(1–2). Available at www.cios.org/www/ejc/v17n12.htm (accessed 1 July 2015).

Skelton, T. 2009. Children's geographies/geographies of children: play, work, mobilities and migration. *Geography Compass*, 3(4), pp. 1430–48.

Strate, L. 2004. A media ecology review. *Communication Research Trends, Centre for the Study of Communication and Culture*, 23(2), pp. 1–48.

Symes, C. 2007. Coaching and training: an ethnography of student commuting on Sydney's urban trains, *Mobilities*, 2(3), pp. 443–61.

Thomson, L. 2009. *'How times have changed': active transport literature review*. Melbourne: VicHealth.

Tranter, P. and Whitelegg, J. 1994. Children's travel behaviours in Canberra: car-dependent lifestyles in a low-density city. *Journal of Transport Geography*, 2(4), pp. 265–73.

Valentine, G. 1997. 'Oh yes I can' 'oh no you can't': Children and parents' understandings of kids' competence to negotiate public space safely. *Antipode*, 29(1), pp. 65–89.

Valentine, G. and McKendrick, J. 1997. Children's outdoor play: exploring parental concerns about children's safety and the changing nature of childhood. *Geoforum*, 28(2), pp. 219–35.

Verbeek, P. 2005. *What things do: philosophical reflections on technology, agency and design*. University Park, PA: Pennsylvania State University Press.

Wajcman, J. 2008. Life in the fast lane? Towards a sociology of technology and time. *The British Journal of Sociology*, 59(1), pp. 59–77.

Wajcman, J., Bitman, M., Jones, P., Jonhnstone, L. and Brown, J. 2007. *The impact of the mobile phone on work/life balance. Preliminary report*. Canberra, Australian National University: Australian Research Council Linkage Project.

Whitzman, C., Worthington, M. and Mizrachi, D. 2009. *Walking the walk: can child-friendly cities promote children's independent mobility?* Melbourne: Australasian Centre for the Governance and Management of Urban Transportation (GAMUT).

Williams, S. and Williams, L. 2005. Space invaders: the negotiation of boundaries through the mobile phone. *The Sociological Review*, 53(2), pp. 314–31.
Zubrick, S. R., Wood, L., Villanueva, K., Wood, G., Giles-Corti, B. and Christian, H. 2010. *Nothing but fear itself: parental fear as a determinant of child physical activity and independent mobility*. Melbourne: Victorian Health Promotion Foundation (VicHealth).

8   The role of the geography of educational opportunity in the wellbeing of African American children

K. Milam Brooks and Pamela Anne Quiroz

The notion of 'geography of opportunity' is rooted in the idea that where people live affects their access to academic and economic opportunities, life outcomes and overall wellbeing (e.g. health, life span and happiness). In other words, the place where you live has a powerful role in the quality of the education received.

Research finds a negative effect on academic achievement for those children who attend a high poverty school (Acevedo-Garcia *et al.*, 2008; Briggs *et al.*, 2008). Children living in dense, low socio-economic urban communities are often exposed to a number of risk factors that threaten educational wellbeing. Negative educational effects of low performing schools are further fuelled by inadequate access to health and human services, limited economic opportunities and high levels of violence and crime (Leventhal and Brooks-Gunn, 2004). Even in the most disadvantaged communities, however, schools have an obligation in promoting wellbeing and in detecting when a young person's wellbeing may be at risk.

The expansion of neo-liberal influence on the restructuring of public education, particularly in urban areas, has promoted market oriented approaches and a belief that parents, and hence, their children are consumers of education, and these educational alternatives now include both public and private schools.

This chapter explores the impact of the geography of educational opportunity on the wellbeing of low-income African American fourth graders (9–10-year-olds) who attended a newly developed private school in an affluent part of Central City (pseudonym),[1] where wellbeing was defined as safety, academic engagement and also positive racial identity.

US school reform and the World Citizens School

Education in the US is undergoing a transformation and there is considerable parent dissatisfaction with current educational policies, such as high-stake standardised testing and zero-tolerance discipline policies that seem to encourage student disengagement, particularly by African American children. One of the many transformations taking place in US schools involves the geography of educational opportunity because school 'choice' now dominates reforms to enhance opportunities to obtain a quality education. As is also common in other countries, parents send their children to the 'best' school (e.g. Lewis, 2004; Kraftl, 2013). Although where a family lives is associated with the quality of education their children receive, education is still the primary mechanism available to obtain social mobility and opportunities to attend school have now expanded beyond neighbourhood public schools to other educational alternatives. Galster and Killen (1995) argue that individuals' lives can be profoundly changed if they move to an environment that offers new opportunity. Ethnically and culturally under-represented groups and working class families are increasingly seeking alternatives to neighbourhood schools. As the landscape of education changes it is important to examine how alternative educational spaces (e.g. Mills and Kraftl, 2014) and practices impact children's wellbeing. The World Citizens School (WCS) is an example of one such alternative.

The WCS was established in 2006 as a private school as a response to disgruntled African American and Latino parents of Westside, one of Central City's poorest neighbourhoods. The school began as a day-care centre and evolved to an elementary school, adding a grade each year. As the WCS expanded, it relocated to a much larger space in Riverwood, one of the city's wealthiest neighbourhoods. Riverwood is located at the heart of Central City with a plethora of material, political, cultural and economic resources. At the time of our study, the first cohort of students at the WCS were in fourth grade. Combining a rigorous academic structure, a 'grow-your-own' philosophy and pro-cultural racial socialisation, the WCS appeared to be promoting academic engagement and confident students who were learning about the global environment in which they live. Pro-cultural racial socialisation refers to the philosophy that social and cultural values and knowledge shape and reshape the African American experience (see Winkler, 2012). Nevertheless, and regardless of its apparent success, such educational experiments often generate substantial debate among community members and activists – and the WCS was no exception. Critics questioned the need to locate the school outside the Westside community, as well as the implicit assumption that cultural capital could only be accessed from wealthy (and predominantly white ethnic) neighbourhoods (Orfield and Frankenberg, 2012; Wells, 2008). This example contrasts the approach of the Harlem Children's Zone, which takes education into the heart of a deprived community (Harlem Children's Zone, 2015). Others questioned the school's selection process and 'creaming of the crop', arguing that such educational alternatives further reduce opportunities for the development of social capital among community members who are left behind. In essence, the development of the school and the founders' goals to eventually build dormitories reminiscent of New Zealand youth health camps (see Kearns and Collins, 2000) and move children from the Westside to the WCS campus raised questions about whether children should be required to leave their community to access quality education and how changing the geography of educational opportunity contributed to children's overall wellbeing.

At the time of the study (January–June 2011), the WCS had an enrolment of 125 students that included two kindergarten classrooms and also one classroom each for first through to fourth grade. Although priority was given to children who began at the WCS day-care centre, all applicants were required to apply and take an entrance exam.[2] The school was 87 per cent African American, 10 per cent Latino, 2 per cent Asian and 1 per cent Caucasian (see Table 8.1). According to school data, 90 per cent of students were low-income and qualified for the US government free or reduced lunch program. The cost of tuition was approximately US$7,000 per year and staff estimated that 95 per cent of students attended the school on scholarships provided through government funds, private benefactors and fundraising activities.

The WCS utilised Direct Instruction (DI) as the primary mode of instruction. Created by Engelmann and Becker in the 1960s, DI is a model of instruction that focuses on well-developed and carefully planned lessons designed around small learning increments, along with clearly defined and prescribed tasks (Engelmann, 2007). The school coupled this form of instruction with its regular use of standardised tests to help assess student progress, improve their skill sets and assure that students would be prepared to achieve in other school environments. DI creators believe that all children, particularly disadvantaged learners, can improve academic skill sets and self-image. These principles are consistent with the philosophy of the WCS. Another distinguishing feature of the WCS was its attention to pro-cultural racial socialisation, a central and intentional feature of promoting African American children's wellbeing. Central City has long been one of the most racially segregated cities in the US, and more than 50 per cent of its children attend what the Central City school district used to call 'racially isolated' schools (i.e. when more than 85 per cent of the students attending a given school are of the same race or ethnicity). Nevertheless, the vast majority of these same-race schools fail to attend to the pro-cultural racial socialisation of their students. At the WCS, learning involved rituals steeped in tradition where role models promoted the inherent value of African American heritage because the children were immersed in a curriculum that included the histories, current events and contributions from African and African American cultures.

Table 8.1 Demographic details for the World Citizens School and its neighbourhoods

	Westside	*Riverwood*
Median income	$22,426	$75, 248
College degrees (%)	5.6	64.9
Professional careers (%)	18	76.5
Racial/Ethnic composition		
Black (%)	69.2	9.92
Latino (%)	26.4	19.7
Asian (%)	0.4	5.95
White (%)	2.4	62.1
Other (%)	1.6	2.33

Source: US Census Bureau, n.d.

Methods

We collected data between January and June 2011 in the form of school documents, interviews, observations, social media and child-centred activities. For the purposes of this paper, we examined these data to assess wellbeing qualitatively as defined by the school: physical, academic and racial wellbeing.

We observed students in and out of school to understand the spaces where they live and learn. A walking audit of the Westside gave us a sense of how the landscape of the neighbourhood contributed to children's physical and psycho-social wellbeing, which is typically operationalised as 'safety'. Added to this were children's photographs and community maps that allowed children to visually express the features of their Westside neighbourhood that mattered to them and provide a glimpse into their everyday lives. Children were asked to draw community maps of their neighbourhood including streets, houses, parks and other local spaces, along with their favourite places to hang out. The guided map activity occurred over a two-week period and served as a starting point for exploring children's lives from a visual perspective. They also engaged in self-directed photography to augment their community maps.

Academic wellbeing was assessed using observations both in and out of the school. These included observations of the fourth-grade classroom and other classroom activities, school assemblies, recess, lunch and field trips. We observed staff meetings, parent–teacher conferences and parent committee meetings. We rode the bus to and from school with the children and attended school performances or watched them on social media.

Although somewhat murkier, racial wellbeing was indicated when children exhibited knowledge and pride during performances, classroom interactions and activities that explicitly focused on racial contributions, knowledge and skills, such as Tabura – a form of Brazilian martial arts. In-class knowledge about racial identity, history, social issues related to race, and other formal and informal comments about race were assessed on a continuum of negative to positive racial views (see also Berg, 1993). We also engaged in informal conversations with students and school staff during fieldwork. These data collectively provided a sense of general wellbeing and a glimpse into the specific aspects of wellbeing attended to by the school, and how they contributed to the development of these African American children.

Physical wellbeing

During our walking audit we counted the number of institutional supports (schools, hospitals, clinics, firehouses, grocery stores) in a four-square-mile area. We found minimal green space, numerous store-front churches (often associated with high poverty areas that traditional churches have long ago abandoned), currency exchanges, liquor stores, littered and abandoned lots, and metal grates and bars on store-front windows and the first floors of homes (see Figure 8.1). The physical wellbeing of Westside is at constant risk due to

Figure 8.1 A common street scene in Westside.

community violence. We observed a heavy police presence, either patrolling the neighbourhood in cars or via surveillance cameras mounted at the top of street light poles. These surveillance cameras are distinguishable by their flashing blue lights, similar to those found on police vehicles. Children spoke about the constant sound of police sirens as an indicator of something bad happening and this contributed to perceptions of feeling unsafe. Children's photographs showed very little or no green space or parks and more dilapidated buildings and empty lots. Most striking was that most of the children took the photographs of outdoor space from either the window of their home or the window of the school bus or other vehicle (see Figure 8.2). Photographs that were taken outside of the home indicated that children were standing directly outside their front door. These visual indicators offer a glimpse of how the physical environment of the neighbourhood impacted children's sense of safety, a prominent theme in their interviews.

The school was intentionally located in an affluent neighbourhood to address the issue of safety, access cultural capital and possibly to attract more well-off families who would not require financial assistance from WCS. Students contrasted their lives in their neighbourhood as unsafe with descriptions of life at school as free from altercations and worry. The aesthetic of both the school's design and its surroundings created a striking contrast to that of the Westside neighbourhood. It was easy to comprehend how the phenomenology of students regarding community could shift from home neighbourhood to that of the school because children uniformly described their school as their community and they conveyed having little attachment to the Westside neighbourhood, partly fuelled and desired by long school days. The parents with whom we spoke seemed very satisfied with this outcome and with the school.

Figure 8.2 A child's view on Westside from a car.

> Mr Wilson: I don't know if that type of environment [referring to the city's public schools] would be welcome for him [the son], the type of school where you know, you just have to deal with it . . . I was in a situation like that. I was bullied, I was made fun of . . . I wasn't able to have conversations with my parents or teachers about it, whereas he's talking to us about what's happening to him, happening *with* him, happening *around* him, and I think that the school helps to foster that type of discussion.

Safety is one of the key measures of child wellbeing according to Coleman (1990). Although we were initially surprised by parent satisfaction with the time their children spent away from the neighbourhood and in the school (i.e. up to 12 hours a day), we also realised that for families who reside in communities devastated by poverty, the consequences of cumulative and negative life circumstances can be overwhelming. One of the solutions adopted by most of these parents was to spend as much time with their families outside the neighbourhood as possible. Although a permanent move was not financially viable, they opted for a social retreat from the local community instead. This was one of the primary appeals of the WCS.

Educational wellbeing

The photographs in Figures 8.1 and 8.2 provide a glimpse of the Westside neighbourhood. Because of lack of resources, crime or urban design (e.g. parks and playgrounds), 'protective factors' (e.g. family, church, schools) gain salience for disadvantaged families because they generate resilience among children whose families live in poverty (Benard, 2004; Wridt, 2010). Protective factors, like schools, serve as buffers to risk-producing conditions like poverty. The founders of the WCS were therefore intentional in their design of the school to serve as a protective factor in the lives of its African American students. The school's location, selection of staff and use of uniforms, rituals and daily and extra school activities reflected this design. Having recognised family as the foundational support of a child's wellbeing, the WCS also provided a number of parent workshops. In the classroom, teachers helped to create an environment that enriched the lives of their students, and resilience research suggests that committed and caring teachers play an indispensable role by helping students to develop the ability to triumph over any challenge (Brown *et al.*, 2001; Noddings, 2005; Valenzuela, 2005). One teacher used Twitter feeds from various countries to help children understand the concerns of a country's citizens. The children were also required to read the newspaper daily to keep up with current and world events. We observed the daily newspaper reading and lively conversations about the news during one of our many bus rides. In class children were required to report and write about what was happening in the world based on what they had read in the newspapers, and when the teacher asked questions the expectation was that all children would raise their hand to provide a response. During discussions children appeared to demonstrate critical thinking skills because responses to questions about world issues were carefully crafted and generally sophisticated. These activities were designed to prepare children for international travel where they would learn about other places and cultures through in-country service learning activities. The teachers linked student (and family) experiences to current events in the course of working together to develop pragmatic responses to these issues. In general, classroom observations demonstrated impressive instruction and high expectations to which most students responded. Although standardised tests did not drive instruction, they were used intermittently to help staff assess children's progress and needs.

Observations of fieldtrips to local museums, cultural events and performances also allowed us to gain a sense of children's academic wellbeing. Whether on guided museum tours or performing Capoeira (Brazilian Martial Art) for a university audience, children at the WCS presented themselves as exceptionally able and confident. Indeed, performances of Capoeira and black history were so popular that several colleges and universities requested them – a trend that was parallel to developments in Maori-language immersion schools (Harrison, 2005).

Racial wellbeing

Research has shown the relationship between positive racial identity and wellbeing, and it is indeed important that schools affirm the identities of their students, especially African American students (see Caldwell *et al.*, 2002; Sellers *et al.*, 2003; Hughes *et al.*, 2015), particularly because race continues to structure inequality in the US (Omi and Winant, 1994). This affirmation is implicit as a central component of learning that is integrated into each subject area. For example, when learning about the various styles and metrics of poetry, African and African American poets are used to illustrate lessons. The majority of books in each classroom library, as well as the school's main library, contained African American characters and storylines. These subtle aspects of pro-cultural racial socialisation were simply one feature of children's positive sense of self, but we observed them in all aspects of the schooling process that lent to what Winkler (2012) called a racial safe space, and what Vernon and Papps (2004) referred to as cultural safety. According to Winkler (2012), it is a racial safe space where one can fully express him/herself without the fear of rejection or reprisal. Similarly, cultural safety is met through the actions of wellness professionals who recognise, respect and nurture the cultural identity of those they serve, while meeting their needs and expectations (Wepa, 2004).

The WCS as a 'protective factor'

Schools can play an important role in promoting wellbeing and in detecting when a child's wellbeing may be at risk. Aldgate and McIntosh (2006) find that although schools can be a source of stress for children, they can also be places of opportunity and serve as a protective factor. Students regarded the WCS in this way, as the following comments indicate:

> Damien: This school is fair and they're very nice. We get to learn things about our history that a lot of others schools don't learn.

> Jason: At the other school (referring to a Westside neighbourhood public school) there were lots of fights in the halls and on the playground. It was just bad over there and I was afraid sometimes . . . it's not like that here (referring to the WCS), they care about us.

Although wellbeing in the US is typically associated with independence and autonomy, the WCS appeared to promote a type of wellbeing based on interdependence and the school as community. Children appeared to have developed close bonds and engaged in activities to achieve those bonds because they presented themselves as well-adjusted, confident, high academic achievers. Nevertheless, we realise that this was a self-selected (and school selected) group of children, and those who did not perform as well or whose families did not subscribe to the school's philosophy were likely no longer at the WCS. We also recognise that we achieved merely a snapshot of the children's lives and the processes they were experiencing, making it impossible to assess the long-term impact and wellbeing of these children. Finally, despite the potentially positive outcomes for these children, the question remains about whether removing the children from their home neighbourhood is the best way to achieve wellbeing.

Jeffrey Canada is founder of the Harlem Children's Zone, which shares the same demography as the WCS in terms of racial and socio-economic status. Harlem Children's Zone has been successful in achieving results similar to the WCS without moving the children out of their neighbourhood. Some would argue that this success has occurred by recognising and building on the neighbourhood's available social and cultural capital. The Harlem Children's Zone is a public charter school in Harlem, New York based on the Beacon School Model that combines comprehensive services for students with activities based on the youth development model. In addition, they provide a full network of services from birth to college for a one-hundred-square-block area in Harlem that includes adult education, social services and community building programs. The fundamental principles of the Harlem Children's Zone are to help children as early in their lives as possible and to create a critical mass of adults around them who understand what it takes to help children succeed (Harlem Children's Zone, 2015).

If a school like the Harlem Children's Zone provides a blueprint of how impoverished communities can develop and promote wellbeing in children by retaining a commitment to place instead of substituting place then, we may ask which is the better model? One can argue that because children of the Harlem Children's Zone spend time in their neighbourhood they are more likely to remain connected to the community, and such connections promote possibilities for sustaining and investing in the community as opposed to the disinvestment of parents like those at WCS. Another critique of the WCS is that in addition to the political and economic disinvestment of cities in particular neighbourhoods, student migration results in the attenuated ties of families to neighbourhood and it limits the development of human, social and cultural capital, a phenomenon that sociologist William J. Wilson (2012) labelled the 'truly disadvantaged'. The flight of those who 'can' leaves neighbourhoods like Westside devoid of the talent and resources of its members. Such neighbourhoods benefit immensely when new forms of capital and resources are vested in the community, helping not only to sustain but also to regenerate them.

Conclusion

Kraftl (2013) challenges us to carefully and critically examine educational exper- iments like the WCS to see how or if they challenge the neo-liberal based policy that seeks to undermine public education through market-driven models and school choice. We know from previous research that growing up in poverty impacts all areas of a child's life through multiple and cumulative risks that affect their overall wellbeing. Since children spend a significant portion of their lives in schools, we need to understand how and what role schools play in promoting wellbeing for children. We notice that the physical and academic environment and the positive messages that children at the WCS receive about who they are enhance their ability to achieve and support their overall wellbeing. But what are the consequences for children who spend so much time outside their home commu- nity? Because these children lived in a neighbourhood where safety was a major concern, their parents made sure that time out of school was spent in structured activities outside of the neighbourhood. This made it very difficult for the children to form friendships in the neighbourhood. Close community ties are a measure of wellbeing and the long-term impact of educational youth mobility therefore remains unclear. We learned about the close bonds children had with the WCS, with many referring to the members of the school as family. Nevertheless, the Harlem Children's Zone provides an alternative example of how wellbeing can be achieved without altering place and by affirming that communities, even poor African American communities, have cultural capital. Is it therefore necessary or even optimal to move outside the neighbourhood in order to achieve wellbeing?

As we contemplate how educational opportunity can be expanded to those communities that need it the most, we must also address the complicated and often conflicting messages and outcomes of educational experiments like the WCS. To impede student migration may deprive children of accessing quality educational opportunities and life-altering experiences. Moreover, a protective environment that also serves as a racial safe space nurtures overall wellbeing. Although the physical and academic wellbeing of children, rather than risks related to race, is more easily assessed through tools like the Although the physical and academic wellbeing of children is more easily assessed through measurement tools than risks related to race, we must infer the positive values of a pro-cultural racial socialisation through our glimpse of the children at the WCS. These issues and the impact they have on children are obviously tied to how we define and prioritise not only wellbeing but also the wellbeing of African American children.

Notes

1 In the US children are considered a vulnerable population and research must therefore protect their identity as much as possible. Pseudonyms have therefore been used for school names, cities, neighbourhoods and participants.
2 Children are accepted into the WCS day care through an open enrolment process. Acceptance criteria were considered confidential and not provided to the researchers.

References

Acevedo-Garcia, D., Osypuk, T.L., McArdle, N. and Williams, D.R. 2008. Toward a policy-relevant analysis of geographic and racial/ethnic disparities in child health. *Health Affair*, 27(2), pp. 321–33.

Aldgate, J. and McIntosh, J. 2006. *Time well spent: a study of wellbeing and children's daily activities*. Edinburgh: Astron.

Benard, B. 2004. *Resiliency: what we have learned*. San Francisco, CA: WestEd.

Berg, L. 1993. Racialisation in academic discourse. *Urban Geography*, 14, pp. 194–200.

Briggs, X., Ferryman, K., Popkin, S. and Rendón, M. 2008. Why didn't the moving opportunity experiment get children to better schools? *Housing Policy Debate*, 19(1), pp. 53–91.

Brown, J.H., D'Emidio-Caston, M. and Benard, B. 2001. *Resilience education*. Thousand Oaks, CA: Corwin Press.

Caldwell, C.H., Zimmerman, M.A., Bernat, D.H., Sellers, R.M., and Notaro, P.C. 2002. Racial identity, maternal support, and psychological distress among African American adolescents. *Child Development*, 73(4), pp. 1322–36.

Coleman, J.S. 1990. How worksite schools and other school reforms can generate social capital: an interview with James Coleman. *American Federation of Teachers*, 14(2), pp. 35–45.

Engelmann, S. 2007. Student-program alignment and teaching to mastery. *Journal of Direct Instruction*, 7(1), pp. 45–66.

Galster, G.C. and Killen, S.P. 1995. The geography of metropolitan opportunity: a reconnaissance and conceptual framework. *Housing Policy Debate*, 6(1), pp. 7–43.

Harlem Children's Zone. 2015. *Harlem Children's Zone: about us*. Available at http://hcz.org/about-us/ (accessed 5 March 2015).

Harrison, B. 2005. The development of an indigenous knowledge program in a New Zealand Maori-language immersion school. *Anthropology and Education Quarterly*, 36(1), pp. 57–72.

Hughes, M., Kiecolt, K.J., Keith, V.M. and Demo, D.H. 2015. Racial identity and wellbeing among African Americans. *Social Psychology Quarterly*, 78(1), pp. 25–48.

Kearns, R.A. and Collins, D.C. 2000. New Zealand children's health camps: therapeutic landscapes meet the contract state. *Social Science and Medicine*, 51(7), pp. 1047–59.

Kraftl, P. 2013. *Geographies of alternative education: diverse learning spaces for children and young people*. Bristol: Policy Press.

Leventhal, T. and Brooks-Gunn, J. 2004. A randomized study of neighbourhood effects on low-income children's educational outcomes. *Developmental Psychology*, 40(4), pp. 488–507.

Lewis, N. 2004. Embedding the reforms in New Zealand schooling: after neo-liberalism? *GeoJournal*, 59, pp. 149–60.

Mills, S. and Kraftl, P. eds. 2014. *Informal education, childhood and youth: geographies, histories, practices*. Basingstoke: Palgrave Macmillan.

Noddings, N. 2005. *The challenge to care in schools: an alternative approach to education*. New York: Teacher College Press.

Omi, M. and Winant, H. 1994. *Racial formation in the United States: from the 1960s to the 1990s*. New York: Routledge.

Orfield, G. and Frankenberg, E. 2012. *Educational delusions? Why choice can deepen inequality and how to make schools fair*. Berkeley, CA: University of California Press.

Sellers, R.M. and Shelton, J.H. 2003. The role of racial identity in perceived racial discrimination. *Journal of Personality and Social Psychology*, 84(5), pp. 1079–92.

US Census Bureau. n.d. *2010 Census Data*. Available at https://www.census.gov/2010census/data/ (accessed 28 February 2017).

Valenzuela, A. 2005. Subtractive schooling, caring relations, and social capital in the schooling of US–Mexican youth. In: L. Weis and M. Fine, eds. *Beyond silenced voices: class, race, and gender in United States schools*. Albany, NY: State University of New York Press.

Vernon, R. and Papps, E. 2004. Cultural safety and continuing competence. In: D. Wepa, ed. *Cultural safety in Aotearoa New Zealand*. Cambridge: Cambridge University Press.

Wells, A.S. 2008. The social context of charter schools: the changing nature of poverty and what it means for American education. In: M. Springer, H. Walberg, M. Berends and D. Ballou, eds. *Handbook of research on school choice*. Philadelphia, PA: Lawrence Erlbaum Associates.

Wepa, D. (ed.) 2004. *Cultural safety in Aotearoa New Zealand*. Cambridge: Cambridge University Press.

Wilson, W.J. 2012. *The truly disadvantaged: the inner city, the underclass, and public policy*. Chicago, IL: University of Chicago Press.

Winkler, E. 2012. *Learning race, learning place: shaping racial identities and ideas in African American childhoods*. New York: Rutgers University Press.

Wridt, P. 2010. A qualitative GIS approach to mapping urban neighbourhoods with children to promote physical activity and child-friendly community planning. *Environment and Planning B: Planning & Design*, 37(1), pp. 129–47.

9 Child medical travel in Argentina

Narratives of family separation and moving away from home

Cecilia Vindrola-Padros and Eugenia Brage

Childhood is a unique stage of the life-course. How childhood is conceptualised locally shapes and is shaped by ideas on dependency, autonomy, adulthood and obligations, as well as micro-level power relations. These factors determine children's quality of life and their participation and agency in everyday urban life (Carroll *et al.*, 2015). But what happens when children fall ill and need to seek treatment? What happens when children need to leave their local family and friendship networks to be treated in a different city far away from home?

Studying children's medical travel gives us a direct glimpse into the strategies children and their families use to obtain health services without losing sight of the ways in which children's health and travel experiences are shaped by economic, social, cultural and political factors. It also points to a different way of experiencing childhood: on the move, away from home, grappling with a disease and coping with medical treatment.

In this chapter we present the results of a study on the medical travel experiences of children seeking oncology treatment in Argentina. The purpose of the chapter is to highlight the value of examining medical travel through the eyes of children and their parents and therefore understand the role played by the disease, medical treatment and relocation to a new city in their own terms. This focus on the micro-level aspects of child medical travel allows us to emphasise the complex and multidimensional nature of the process of seeking medical treatment in a different locale.

It is estimated that between 60,000 and 85,000 adults and children travel worldwide each year to obtain medical attention (Ehrbeck *et al.*, 2008). This figure only includes international migrants and excludes the large number of people who travel within their country's borders to access care. The number of 'mobile patients' is therefore exponentially higher, and because of the unequal distribution of medical resources across the globe, it will likely continue to rise.

The concept of medical travel has been proposed by social scientists to highlight the complexities of journeys and the wide range of movement strategies implemented by people seeking medical attention (Sobo, 2009). This rapidly growing area of research has focused on identifying the motivations behind seeking travel in another location, indicating that decision-making processes are complex and that most people relocate to seek treatment not available in their place of origin, obtain better quality or quicker care, access novel or experimental treatment, and acquire services in contexts more amenable to their cultural preferences (Kangas, 2007; Song, 2010; Whittaker and Speier, 2010). Research has focused on the burden of travel on the health care systems in host countries (Whittaker, 2008; Whittaker *et al.*, 2010) and the opportunities and hardships that travel presents for people requiring care and their family members (Kangas, 2007; Solomon, 2011).

Even though research on the travel experiences of adults seeking medical attention has notably increased over the past decade (Kangas, 2007; Sobo, 2009; Lee *et al.*, 2010), studies on children who must relocate to access complex medical treatments are scarce. The few available studies have explored child medical travel from either the perspective of health care professionals (Senior, 2006; Massimo *et al.*, 2008; Culley *et al.*, 2013) or the child's parents (Crom, 1995; Vindrola-Padros and Whiteford, 2012; Margolis *et al.*, 2013). These studies have pointed to the challenges faced by parents and healthcare professionals due to language barriers, different cultural backgrounds and high treatment costs.

However, missing from the literature are the experiences of the large number of children who require medical treatment – lifesaving in many cases – for complex diseases such as cancer. We know very little about the conditions under which they travel, the arrangements made by families to pay and facilitate their relocation and the emotional impact, not only of the diagnosis, but also of having to leave their home and family behind. In the following sections, we will shed some light on the complex medical and emotional negotiations families make when seeking medical treatment.

The complexity of child medical travel

Studies on mobility and health have explored the relationship between individuals' lack of access to health care in their place of origin and their use of mobility to obtain desired services (Ellis and Muschkin, 1996; Wood *et al.*, 2000; Ensor and Cooper, 2004; London *et al.*, 2004; Cham *et al.*, 2005; Elmore, 2006; Gutierrez, 2008; Molesworth, 2005; Asiedu Owusu and Amoako-Sakyi, 2011). Healthcare access is seen as a complex process, often involving multiple barriers and requiring different types of mobility strategies. People might need to travel to access care elsewhere because of the lack of physical medical facilities and personnel, an inability to physically access those facilities (for instance, unsuitable roads and means of transport), an inability to afford services in those facilities, or because of perceptions they might have regarding the services and personnel in those facilities (for instance, feelings of distrust, ideas that the quality of care is not adequate, or previous negative experiences) (Penchansky and Thomas, 1981; Ricketts, 2009; Ergler *et al.*, 2011).

An understanding of people's perceptions of the health system is an important and often overlooked aspect of studies on healthcare access. As Ergler *et al.* (2011: 336) have indicated, 'feelings can influence how care is both received and delivered, as well as which discourses on the part of both service providers and users are reinforced'. Ideas people might have about the health system, types of services and quality of staff can influence their health-seeking behaviour and determine their decision to seek care in a different locality (Ergler *et al.*, 2011). Access to care is not only determined by proximity to services or the patient's ability to afford and secure travel options, it is also influenced by patients' attitudes towards healthcare establishments and personnel, their satisfaction with care, comfort with their surroundings and the characteristics of the care providers (Neuwelt *et al.,* 2015). Healthcare access needs to be seen as a highly complex process, involving multiple geographical, social, cultural and political scales (Borges Guimarães, 2007).

When the desired care is not available close to home, mobility becomes a tangible representation of individual agency where, in light of barriers to care, individuals engage in the process of leaving their place of origin, searching for a new locale and relocating to a destination where desired services are available (Porter, 2007; Porter and Hampshire, 2010; Hampshire *et al.*, 2011). Most of the current research examining the relocation of patients seeking care to a region outside of their place of origin has used terms such as 'medical tourism' and 'medical travel' (Connell, 2006, 2013; Hopkins *et al.*, 2010; Smith *et al.*, 2011). In these studies, travel often implies a specific trip with defined purposes. Our in-depth research on the experiences of medical travellers paints a different picture – one where children seeking care and their families are engaged in multiple 'degrees' of travel (intra-city, inter-city, regional, national, international) and sometimes even find themselves in a constant state of movement.

The child requiring care might travel to a nearby facility for some aspects of care and then relocate temporarily to another for more specialised procedures (Hampshire *et al.*, 2011). They might travel to request appointments, obtain medication, hand in paperwork, attend follow-up appointments or collect study results (Gutierrez, 2008; Hampshire *et al.*, 2011; Johnson and Vindrola-Padros, 2014). Some of their traveling might be arranged by a hospital or charity, while other types might need to be secured by the child requiring care or their family (Vindrola-Padros, 2012). Some travel might be performed with one accompanying family member, while others might involve more companions (Vindrola-Padros, 2011, 2012; Vindrola-Padros and Whiteford, 2012).

In other words, medical travel must be seen as continuous processes entailing journeys of different distances, duration, frequency, complexities and impact (financial, social and emotional) that intertwine to create particular treatment experiences. The preparation and arrangements that need to be put in place for these journeys, as well as children's (and their families') experiences of going through them will depend on their individual characteristics (i.e. age of the child, presence of siblings, family income, etc.) and also the local context where they receive the care (distribution of facilities, support for traveling families, transport resources and routes). Their experiences will also change over time as the children and their families acquire information on the best ways to access care. Through interactions with medical staff and hospital processes, they become experts in the intricacies of the healthcare system and learn how to overcome formal and informal barriers to health services (Ricketts, 2009).

Children traveling for oncology treatment in Argentina

In Argentina cancer is the most common cause of death from a disease in children aged 5–15 years (Moreno and Abriata, 2010) and the second most common cause in those aged 1–4 years (Moreno *et al.*, 2009). The public health system provides care for 80 per cent of all paediatric oncology cases using a centralised model where treatment is mainly available in five hospitals located in Buenos Aires (Dussel *et al.*, 2014). Recent reports have indicated that the prognosis of children with cancer in the country varies in relation to their province of residence, where the best prognosis is reported in children residing in the city of Buenos Aires and the worst is found in the north-western and north-eastern parts of the country (Moreno *et al.*, 2013). These differences in prognosis are associated with the unequal distribution of medical facilities and specialised personnel, which leads to delays in diagnosis and referral, and treatment complications in regions outside of Buenos Aires (Instituto Nacional de Cancer (INC), 2015).

In light of this unequal distribution of medical resources, 40 per cent of all children diagnosed with cancer relocate to another city during some part of their treatment (Moreno *et al.*, 2009). Most children travel with one family member, usually their mother (Vindrola-Padros, 2011). Travel might be arranged through a 'system of medical referral' where the child's relocation is guided through a network of centres, or it might be the result of self-referral where the child and family member have made the decision to leave the place of origin in search of medical attention without having a formal referral made by a healthcare professional.

The lack of specialised services in a significant portion of the country has led to the creation of constant and well-established flows of children seeking care in Buenos Aires (Brage *et al.*, 2013). The State has responded to this situation by supplying government aid in the form of travel subsidies and cheap accommodation in Buenos Aires (Vindrola-Padros, 2012); however, this support has done little to subvert the financial, psychological and emotional impact of family separation and homesickness produced by travelling for medical treatment.

The Argentine centralised model of care therefore remains incomplete in its attempt to provide specialised and integrative medical services because it is not able to guarantee continuity of care for children from all parts of the country. Continuity of care is often compromised due to the lack of trained professionals to carry out follow-ups in the place of origin and the non-existence of formalised shared-care models (Dussel *et al.*, 2014). Aside from the clinical services available in Buenos Aires, travel and subsidised accommodation are the only state-funded resources available to children with cancer in need of specialised care.

Methods

The aim of the study was to document the travel experiences of children, young people (5–17 years) and their families seeking care in Buenos Aires, Argentina. Travel experiences were analysed in relation to: (1) children's access to specialised medical attention; (2) the identification of the different types and levels of travel involved in seeking oncology treatment in Buenos Aires; and (3) the impact of relocation on family relationships, income, education (school attendance) and children's regular activities (recreation, sports, socialising with friends, etc.). Special attention was paid to the meaning children and their families attributed to the disease, medical treatment, the need to travel to access medical services and relocation.

Data collection took place from May 2008 to August 2010 (organised in three periods of fieldwork) in the facilities of Fundación Natalí Dafne Flexer (FNDF), a non-governmental organisation that provides social services, information, medication and other forms of assistance to paediatric oncology patients and their families. Thirty-five families were included in the study. In each family the child and one accompanying parent were interviewed. The children were aged between 5 and 17 years at the time of the interview, but only information relating to younger children (5–12 years) is included in this chapter. The children had a wide range of cancer diagnoses, were in different stages of treatment at the time of the interview and travelled to Buenos Aires from 15 different provinces.

The first author carried out open-ended semi-structured interviews with children and family members on their experiences of care, travel and relocation. Drawing techniques during the interviews were used to explore children's views on different stages of treatment and happy and difficult moments (see also Johnson *et al.,* 2012). These methods were used as elicitation devices to guide the interviews with the children. Documentary analysis of health policies, epidemiological data and government statistics was carried out to contextualise the data collected from the observations and interviews. Each method provided a distinct layer of insight into children's experiences. The drawings were particularly helpful in uncovering the impact of migration on children's everyday lives.

The interviews and notes from the participant observation were transcribed verbatim and anonymised (the names used in this chapter are pseudonyms). The transcripts and visual material produced by the children were analysed for content using Atlas.ti (ATLAS.ti Scientific Software Development, 2006), computer-assisted qualitative data analysis software. The process of analysis involved coding, the construction and analysis of thematic nodes, and the creation of conceptual maps to search for relationships between codes and larger categories of analysis. Ethical approval for the study was granted by the Board of Directors of FNDF and the Institutional Review Board of the University of South Florida.

Different types of travel

Even though each family experienced mobility in a particular way, their movement was mainly motivated by the need to obtain one or more of the following: the right diagnosis, the 'best' treatment/care at the hospital, follow-up care at the hospital, specialised care in case of emergency or relapse, other services (social, financial, etc.), medication, and paperwork and appointments. In this chapter, we limit our discussion to families' experiences of obtaining a diagnosis, receiving medical therapies, travelling within Buenos Aires to secure care and other required services, and accessing follow-up care. We use brief vignettes of the experiences of a selected number of families to illustrate common trends found in most of the families included in this study. These families were selected to reflect variation in terms of geographical distribution and travel experiences and also because their interviews contained detailed information on their journeys.

The right diagnosis

In Figure 9.1, we have mapped out Marta's journey from Salta to Buenos Aires. The process of obtaining her diagnosis was, according to her father, delayed for over a year. The family visited the local hospital first (A) and a larger local hospital (B) on two occasions, but Marta was sent back home without the correct diagnosis. Her symptoms were dismissed as common cold symptoms. It was not until her health deteriorated and she experienced physical symptoms of fainting and haemorrhaging from the nose and mouth that she was referred to a hospital in the capital of her province (C). The hospital in Salta Capital (C) arranged her referral to the hospital in Buenos Aires (D) where she received treatment. She remained with her father for 8 months in Buenos Aires. She still visits the hospital

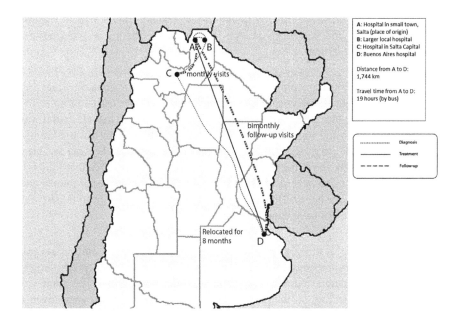

Figure 9.1 Marta's journeys for diagnosis, treatment and follow-up care.

in Salta Capital every month and the one in Buenos Aires every two months. Her father accompanies her and they travel by bus.

A total of ten families encountered problems when attempting to secure a diagnosis for the child. Marta's trajectory in Figure 9.1 is an example of the different types of journeys some families had to experience before receiving a cancer diagnosis, or the suspicion of a cancer diagnosis, and then a referral to a hospital in Buenos Aires. Most families were referred by the local hospital, but others took matters into their own hands and travelled to Buenos Aires on their own.

This was the case of Luisa. Luisa's mother indicated:

> She was pale and they did this analysis saying she had no blood. I went to [local hospital] and they told me it was urgent to get her some blood, but they did not want to give it to her there because they were afraid it was something else. So I came on my own. [. . .] They wanted to take me to Reconquista but I lost too much time with my other child who died and I didn't want the same thing to happen to her.

This was also the case of Juan who was taken out of the local hospital by his parents because he was not being treated and the parents had not received a diagnosis. Care was sought in a private medical facility where a cancer diagnosis was obtained. Juan was then referred by the private facility to a public hospital in Buenos Aires because it was considered to be the best place to obtain treatment.

In the case of Elena, the family did not even attend the local hospital when the child fell ill. They travelled directly to one of the specialised hospitals in Buenos Aires where the child had received care for a bronchiolitis a few years before. Elena's mother explained:

> Since she had the bronchiolitis they asked me if I wanted to continue accessing services for her there [hospital in Buenos Aires]. Before I used to go to my clinic [in the place of origin], where the doctor was seeing her. He said it was bronchospasm and was giving her antibiotics, but she was the same. Monday, Tuesday, Wednesday, she still had a fever. On Sunday I see her and I tell my husband, I think she is purple. 'You think?', he says. It was then that I did not have any money to travel and I told him I would ask my niece for money and then I would take her [to Buenos Aires]. When we got to the hospital she was purple and did not have enough oxygen.

Diagnosis was not a straightforward process for most families and, in several cases, faulty referral processes forced children and their families to endure several trips in search of diagnostic equipment and specialised personnel. The diagnosis routes varied depending on the province of origin. Some families were given a diagnosis in the local hospital, but in all cases, diagnoses needed to be confirmed in Buenos Aires (after additional tests and diagnostic processes).

The 'best' care

Pablo was 7 years old at the time of our interview (we have mapped out his trajectory in Figure 9.2). He started complaining of neck pain and his mother took him to their local hospital (A) three times before one of the doctors decided that he needed to have a magnetic resonance imaging (MRI) scan. The hospital in San Luis did not have this equipment and so he was referred to a hospital in Mendoza (B) where this diagnostic procedure could be performed. In Mendoza he was told he had a brain tumour and needed to undergo surgery. This surgery was originally going to be performed in Mendoza, but the doctors indicated that it would be best to refer him to Buenos Aires where surgeons would have access to better equipment. This referral process took two weeks. The family was referred to a private clinic in Buenos Aires (C) where the surgery and subsequent therapies were carried out over a period of a year and a half. The family returned to their home in San Luis, but Pablo relapsed and needed to be referred to a public hospital in Buenos Aires (D). This referral was made because specialised medical professionals capable of advising and providing better care in cases of relapse were found in Buenos Aires.

The interviews with the children and parents often touched on this idea of 'best' care. Buenos Aires was considered the place with the highest standard of care. Santiago's mother mentioned that other healthcare professionals recommended for her to travel with the ill child to Buenos Aires, 'he was real yellow, his belly was big, blood came out of his nose. He was hospitalised and the doctor would tell me: change doctors, take him to Buenos Aires. They have everything there'.

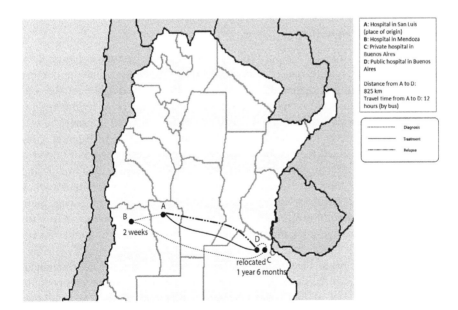

Figure 9.2 Pablo's journeys for diagnosis, treatment and after relapse.

When asked about receiving treatment closer to home, Soledad's mother did not think that was a good idea: 'I don't know, travelling to Buenos Aires, having them look at her here, calms me. I don't really trust that hospital [in the place of origin], I don't trust the doctors there'. Florencia's mother replied in similar fashion saying, 'she can get care in Santa Fe or in Corrientes which is closer, but I know this hospital [Buenos Aires]'.

The 'best' care is not restricted to medical therapies; it has to do with a wider definition of care. When the families talked about services in Buenos Aires they often talked about the 'human side' of the doctors. When asked what she valued most about the services her family had received, Soledad's mother said, 'the human side of the doctors. It is not just that they were treating her, but at the beginning your whole world falls apart when they tell you the diagnosis and the help they give you, they *all* give you'. Martin's mother's comment was similar, 'from the woman who cleans the rooms, to the doctors, the services are not the same [as in the local hospitals]. In my province I did not see it and here [Buenos Aires] the care is spectacular'.

The interviews with the parents thus prompted us to see families' access to health services in a different light because their decisions were not always based on the availability or closeness of the services but rather on what they perceived to be the best type of care for their children. One child, Marta, who was 11 years old at the time of the interview, had a similar viewpoint and did not think receiving all of her care in Salta was a good idea. 'They do not pay as much attention to you there [Salta]. I talk to the doctor and it is like she is not listening to me and she doesn't worry about things'.

Travel within Buenos Aires

Martin had fever for over a month (his trajectory is mapped out in Figure 9.3). His parents took him to the local hospital (A) several times, but the doctors could not give him a diagnosis. When he continued to deteriorate, he was hospitalised in the local hospital. He received a wide range of treatments for four months (mainly antibiotics). After his condition failed to improve, he was referred as an emergency case by plane to a public hospital in Buenos Aires (B). There he received the diagnosis and was treated over a period of one year and two months. Martin and his mother were not able to go back home during his treatment, but they were able to obtain funding from a government agency allowing Martin's brothers and sisters to travel to Buenos Aires to visit them occasionally. Martin and his mother continue to travel to Buenos Aires for his follow-up care every six months and hope to be travelling only once a year soon.

As in Martin's case, all the families in the study stayed in Buenos Aires throughout the entire duration of their treatment. Even in those cases where the family could have received a part of their treatment closer to home, parents and children agreed that it was best to stay in Buenos Aires. A source of concern was what would happen in the case of neutropenia (having a low number of neutrophils and

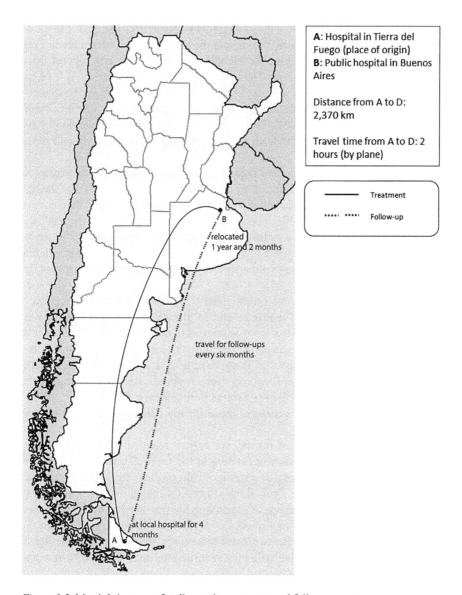

A: Hospital in Tierra del Fuego (place of origin)
B: Public hospital in Buenos Aires

Distance from A to D: 2,370 km

Travel time from A to D: 2 hours (by plane)

———— Treatment

••••• •••• Follow-up

B

relocated 1 year and 2 months

travel for follow-ups every six months

at local hospital for 4 months

A

Figure 9.3 Martin's journeys for diagnosis, treatment and follow-up care.

therefore a higher risk of suffering from an infection) or if another emergency arose. Enrique's mother said:

> From one day to the next I had to leave my house, my family, my baby girl was five at the time and she was raised practically by her grandparents. I was not able to see her for two years. We would talk over the phone, because with the chemo his defence would be down and we could not go.

The healthcare professionals in local hospitals were not seen as capable of dealing with a situation where the life of the child might be put at risk. The best option was to stay in a hotel in Buenos Aires throughout the entire duration of treatment, even if it meant not going back home for months. 'We don't have any support there [place of origin], for her or for us. They treat her like an adult there. [. . .] If I have to stay here [Buenos Aires], I will stay even if it is for the six months of treatment', Maria's mother explained. Sandra's mother had a similar point of view, 'with ambulatory care we still have to be here [Buenos Aires] because we don't have time, if we have to control the fever and are 500 kilometres away, we will never make it in time'.

The hotels where the families stayed during their temporary relocation to Buenos Aires were seen as a home away from home. Marta explained that she had visited several hotels while she remained in Buenos Aires with her father, but the last one was her favourite. 'I like it because it is closer and I know the owners', she said. Felipe decided to draw a picture at the beginning of our interview. When he was finished he said, 'this is where I live. I am on this floor, and this is my window [points to the drawing of a boy with a smile looking out the window], and at the top here it says "hotel"'.

When the children were not hospitalised but families remained in Buenos Aires for the other stages of the treatment, local mobility strategies were used. These trips around Buenos Aires were prominent in children's interviews because many were surprised to see the parks, toy stores and movie theatres in 'the big city'. 'I went to the zoo today. This is my third time. I went with my mom, with my dad and then with both of them together', José indicated. Martin started the interview by saying, 'I like the restaurants here [Buenos Aires] and I can buy videogames. [. . .] I will be getting a guitar for my game today, if it is not too expensive'.

Movement around Buenos Aires was not only recreational, it was also used to hand in paperwork for government funding in the form of travel and relocation subsidies or disability pensions in order to obtain medication from a centralised deposit called the Oncology Drug Bank (Banco Oncológico de Drogas) and to request appointments with medical staff (which often needed to be requested in-person after standing in long queues). Most of the families involved in the study had never travelled to Buenos Aires before, and this type of internal travel forced them to become familiar with the streets and transport routes of a new, and sometimes frightening, city.

Follow-up care

Follow-up care brought new mobility possibilities for families because they could go back to their place of origin temporarily and travel to Buenos Aires regularly for check-ups. Families talked about this as a return to normality, but a normality that was still frequently disrupted by the need to travel to the capital. Even though some families faced hardship trying to juggle life at home with the child's medical care requirements, and often faced financial hardship to pay for further travel and lodging in Buenos Aires, they reiterated that they adhered to the treatment protocol and never missed appointments. Laura's mother explained,

> Once I was late to an appointment but they still saw my child because there had been an accident on the highway and so they saw her anyway. Another time we overslept and ran to catch the bus because we had to leave our house at 3 am. I usually get up at 2 am and at 3 am the bus leaves. It was leaving and we ran but we caught it! I have never missed an appointment.

Daniela's mother talked about how healthcare professionals saw 'families like hers' – mobile families, families outside of Buenos Aires – and she explained her actions as follows,

> We always come before. We are never behind with treatment. [. . .] I was telling this to the doctor because she didn't want to let us go because she was afraid we would miss our appointments. I will not miss the appointments; I know it is for her wellbeing. I get money for the bus fare. If I don't have any [money], I ask the council or the hospital, they give me and we come to the appointment. I will not miss the treatment. Some doctors think that when parents are from outside [of Buenos Aires] they will not come. I tell them that they are wrong about me because I am always one or two days early.

Families' experiences of travel were multidimensional, involving journeys of different lengths, distances and intensities. The need to travel to Buenos Aires and within the capital contributed to the formation of families' identity as 'mobile families' and, in some cases, parents' and children's stories of treatment appear to reflect a constant state of movement. Regardless of the hardship created by the need to leave their place of origin, the parents defended their decision to move to Buenos Aires by arguing that it was the only way they could provide their children with the best care.

The effects of travel and relocation on family life

The long periods away from home were one of the most difficult parts of their treatment experience as Pablo's mother explained,

> The disease and being far away, because we had never been apart. I got here on the 25th of June and left them [the other children] with their grandmother and they didn't know where I had gone or what had happened until last year when they saw me.

Marta, who always travelled to Buenos Aires with her father, said she missed her mother the most:

> She said that next time she would come with me, but it is because of my little brother that she does not come. He is really close to her and does not want to be apart from her, he cries if she is not there.

Children talked about the hospital and hotel as second homes, but even though Buenos Aires was new and exciting, all children mentioned missing their families and activities in their place of origin. 'When I am not sick, I like to play football. I am the goalie', Sebastian indicated. School was a recurrent topic and many children were concerned that they had to interrupt their education because of the treatment. The hospital school system in Buenos Aires was not fully developed at the time of this study and many children fell behind one or two years.

The return to normality during the follow-up stage was not always an easy transition because children and their accompanying parent returned to families who had had to make different arrangements to cope with their absence. Extended family members often stepped in to care for the children who remained at home and family income was tightened to cover travel and relocation expenses. Martin's mother burst into tears when she talked about her return home, 'the routine was not the same. My children were raised with their father, their grandmother, and others. I did not see them for one year and two months, because I was not able to travel'.

Even though some families were able to get access to government subsidies for travel and lodging in Buenos Aires, most still had to search for additional income to cover expenses. They borrowed money from friends and family, held events within their communities to obtain donations from neighbours, took on additional jobs or sought to work more hours, and sold their cars, televisions or other goods. Enrique's mother explained, 'the only thing we have left is our house because I had to sell everything. I sold my car, a Renault 6, a small bike. He [her husband] works in construction and he sold all of his tools'. The children were aware of the financial hardship suffered by their families. 'He [father] just started working again. He left his job because of me', Marta indicated, 'my older brother is now able to cover him when we come to Buenos Aires for my check-ups'.

Conclusion

The experiences of the families included in this study indicate that even though all children and parents were involved in medical travel, the actual duration, frequency, purpose and effects of travel varied. The mapping of the trajectories of each family point to the need to abandon conceptualisations of medical travel as a linear process (involving solely the place of origin and destination) and instead promote a depiction of this type of mobility as a complex, dynamic and often conflictive process.

Travel for diagnosis was often quick, without notice and full of uncertainty because of the delicate health condition of most children. The experience was complicated even further for families who experienced misdiagnoses, bureaucratic delays or unsuitable referrals. Treatment could appear a more static experience if families remained in Buenos Aires for a prolonged period of time, but they were still mobile within the city, traveling to the hospital for ambulatory care, the drug bank for medication, different government institutions to hand in paperwork, or around the city for recreational purposes. Follow-up care created the opportunity of going back home but still demanded regular travel to Buenos Aires.

Many of the families in this study could have obtained paediatric oncology treatment closer to their home but chose to travel hundreds of kilometres to Buenos Aires to access what they considered to be the best medical services in the country. Medical professionals in Buenos Aires were depicted as highly skilled and caring, while clinical teams in regional hospitals were not trusted.

Regardless of the complications generated by the need to travel and relocate to Buenos Aires, all the families included in the study complied with treatment and were cognisant of the need to adhere to the protocol established by the child's clinical team. Even though appointments were not missed and treatment schedules were not suspended, families still had to endure the financial, emotional and social consequences created by leaving their home and community and experiencing family separation for months. The parents highlighted the issues they faced arranging childcare for the children who remained at home and searching for additional income to cover travel and relocation expenses. The children also talked about the long time they had to spend far away from their family and friends, and brought up concerns about the interruption of their schooling and extra-curricular activities.

To conclude, by looking at medical travel experiences through the eyes of children and their parents we were able to see that the process of seeking medical services in another location is complex, heterogeneous and dependent on the stage of treatment. In Argentina, migration is part of the therapeutic itineraries of people with cancer as nearly half of all diagnosed children migrate at some point during their treatment. The structural factors that determine the options, choices and access to health services and treatments, along with the social, economic, cultural capital of each family shape the quality of life of children and their families.

Although these paths are extremely varied and heterogeneous, most families see them as a 'pursuit of the best quality of diagnosis and treatment'. This search for the 'best' type of care leads us to question distance-based models often implemented to guarantee access to health services because families' decisions to access services are not always determined by proximity. This situation has practical implications for future attempts to decentralise care. Models focused on a geographical redistribution of services should not only seek to provide specialised services (such as cancer therapies) in a wider range of facilities, they should also consider the need to change the negative reputation afforded to regional hospitals.

References

Asiedu Owusu, S. and Amoako-Sakyi, R. O. 2011. Mobility and economic constraints as key barriers to children's health seeking in Ghana. *Social Biology and Human Affairs*, 76, pp. 91–105.

ATLAS.ti Scientific Software Development GmbH. 2006. ATLAS.ti (Version 5.2) [Computer software]. Berlin: Author.

Borges Guimarães, R. 2007. Brazilian health regions and scales: a geographical perspective. *New Zealand Geographer*, 63, pp. 97–105.

Brage, E., Dussel, V., Bevilacqua, M. S., Requena, M. L., Bravo, A., Jerez, C., Uzal, L. and Urtasun, M. 2013. Problemas asociados a tratar niños con cáncer en el final de la vida en casos de residencia alejada de las instituciones tratantes. *VIII Congreso Nacional de Medicina y Cuidados Paliativos*, 14–16 November, Mar del Plata, Argentina.

Carroll, P., Witten, K., Kearns, R. and Donovan, P. 2015. Kids in the city: children's use and experiences of urban neighbourhoods in Auckland, New Zealand. *Journal of Urban Design*, 20(4), pp. 417–36.

Cham, M., Sundby, J. and Vangen, S. 2005. Maternal mortality in the rural Gambia: a qualitative study on access to emergency obstetric care. *Reproductive Health*, 2(3), pp. 1–8.

Connell, J. 2006. Medical tourism: sea, sun, sand, and . . . surgery. *Tourism Management*, 27, pp. 1093–100.

Connell, J. 2013. Contemporary medical tourism: conceptualisation, culture and commodification. *Tourism Management*, 34, pp. 1–13.

Crom, D. 1995. The experience of South American mothers who have a child being treated for malignancy in the United States. *Journal of Pediatric Oncology Nursing*, 12, pp. 104–12.

Culley, L., Hudson, N., Baldwin, K. and Lakhanpaul, M. 2013. Children travelling for treatment: what we don't know. *Archives of Disease in Childhood*, 98, pp. 442–44.

Dussel, V., Bevilacqua, M. S., Brage, E., Requena, M. L., Bravo, A., Jerez, C., Uzal, L., Urtasun, M. and Largomarsino, E. 2014. Prácticas y recursos utilizados en la provisión de cuidados paliativos a niños con cáncer en la Argentina. Mapeo nacional Cuidados Paliativos Pediátricos, Informe Ejecutivo. Buenos Aires, Argentina: Instituto Nacional del Cáncer, Ministerio de Salud de la Nación.

Ehrbeck, T., Guevara, C. and Mango, P. D. 2008. Mapping the market for medical travel. Available at http://ww.medretreat.com/templates/UserFiles/Documents/McKinsey%20 Report%20Medical%20Travel.pdf (accessed 22 February 2017).

Ellis, M. and Muschkin, C. 1996. Migration of persons with AIDS – a search for support from elderly parents? *Social Science and Medicine*, 43, pp. 1109–18.

Elmore, K. 2006. The migratory experiences of people with HIV/AIDS (PWHA) in Wilmington, North Carolina. *Health and Place*, 12, pp. 570–9.

Ensor, T. and Cooper, S. 2004. Overcoming barriers to health service access: influencing the demand side. *Health Policy and Planning*, 19, pp. 69–79.

Ergler, C. R., Sakdapolrak, P., Bohle, H. and Kearns, R. A. 2011. Entitlements to health care: why is there a preference for private facilities among poorer residents of Chennai, India? *Social Science and Medicine*, 72, pp. 327–37.

Gutierrez, A. 2008. Geografia, transporte y movilidad. *Espacios*, 37, pp. 100–7.

Hampshire, K. R., Porter, G., Asiedu Owusu, S., Tanle, A. and Abane, A. 2011. Out of the reach of children? Young people's health-seeking practices and agency in Africa's newly-emerging therapeutic landscapes. *Social Science and Medicine*, 73, pp. 702–10.

Hopkins, L., Labonte, R., Runnels, V. and Packer, C. 2010. Medical tourism today: what is the state of existing knowledge? *Journal of Public Health Policy*, 31(2), pp. 185–98.

Instituto Nacional de Cancer (INC). 2015. *Cancer infantil*. Available at www.msal.gov.ar/ inc/index.php/acerca-del-cancer/cancer-infantil (accessed 27 February 2015).

Johnson, G. A. and Vindrola-Padros, C. 2014. 'It's for the best': child movement in search of health in Njabini, Kenya. *Children's Geographies*, 12(2), pp. 219–31.

Johnson, G. A., Pfister, A. E. and Vindrola-Padros, C. 2012. Drawings, photos, and performances: using visual methods with children. *Visual Anthropology Review*, 28, pp. 164–78.

Kangas, B. 2007. Hope from abroad in the international medical travel of Yemeni patients. *Anthropology and Medicine*, 14, pp. 293–305.

Lee, J. Y., Kearns, R. A. and Friesen, W. 2010. Seeking affective health care: Korean immigrants' use of homeland medical services. *Health and Place*, 16, pp. 108–15.

London, A. S., Wilmoth, J. and Fleishman, J. 2004. Moving for care: findings from the US HIV cost and services utilization study. *AIDS Care*, 16, pp. 858–75.

Margolis, R., Ludi, E. and Wiener, L. 2013. International adaptation: psychosocial and parenting experiences of caregivers who travel to the United States to obtain acute medical care for their seriously ill child. *Social Work in Health Care*, 52, pp. 669–83.

Massimo, L. M., Wiley, T. J. and Caprino, D. 2008. Health emigration: a challenge in paediatric oncology. *Journal of Child Health Care*, 12, pp. 106–15.

Molesworth, K. 2005. *Mobility and health: the impact of transport provision on direct and indirect determinants of access to health services*. Basel: Swiss Tropical Institute.

Moreno, F. and Abriata, M. G. 2010. Cáncer en la población de menores de 15 años en Argentina. *Revista Argentina de Salud Pública*, 1(3), pp. 42–5.

Moreno, F., Dussel, V., Abriata, G., Loria, D., and Orellana, L. 2013. *Registro Oncopediátrico Hospitalario Argentino (ROHA)*. Buenos Aires: Ministerio de Salud.

Moreno, F., Schvartzman, E., Scopinaro, M., Diez, B., Garcia Lombardi, M., Loria, D., de Davila, M. T., Kumcher, I. and Goldman, J. 2009. *Registro Oncopediátrico Hospitalario Argentino (ROHA) Resultados 2000–2008*. Buenos Aires: Ministerio de Salud.

Neuwelt, P. M., Kearns, R. A. and Browne, A. J. 2015. The place of receptionists in access to primary care: challenges in the space between community and consultation. *Social Science and Medicine*, 133, pp. 287–95.

Penchansky, R. and Thomas, W. 1981. The concept of access: definition and relationship to consumer satisfaction. *Medical Care*, 14(2), pp. 127–40.

Porter, G. 2007. *Transport, (im)mobility and spatial poverty traps: issues for rural women and girl children in sub-Saharan Africa*. A paper prepared for the international workshop 'Understanding and addressing spatial poverty traps', 29 March, Spier Estate, Stellenbosch, South Africa.

Porter, G. and Hampshire, K. 2010. A moving issue: children and young people's transport and mobility constraints in Africa. *International Forum for Rural Transport and Development Forum News*, 15, pp. 1–3.

Ricketts, T. C. 2009. Accessing health care. In: T. Brown, S. McLafferty and G. Moon, eds. *A companion to health and medical geography*. Oxford: Wiley-Blackwell, pp. 521–39.

Senior, K. 2006. Health migration and childhood cancer. *Lancet*, 7, p. 889.

Smith, R., Martinez Alvarez, M. and Chanda, R. 2011. Medical tourism: a review of the literature and analysis of a role for bi-lateral trade. *Health Policy*, 103, pp. 276–82.

Sobo, E. 2009. Medical travel: what it means, why it matters. *Medical Anthropology*, 28, pp. 326–35.

Solomon, H. 2011. Affective journeys: the emotional structuring of medical tourism in India. *Anthropology and Medicine*, 18, pp. 105–18.

Song, P. 2010. Biotech pilgrims and the transnational quest for stem cell cures. *Medical Anthropology*, 29, pp. 384–402.

Vindrola-Padros, C. 2011. *Life and death journeys: medical travel, cancer and children in Argentina*. PhD thesis, Department of Anthropology, University of South Florida.

Vindrola-Padros, C. 2012. The everyday lives of children with cancer in Argentina: going beyond the disease and treatment. *Children and Society*, 26, pp. 430–42.

Vindrola-Padros, C. and Whiteford, L. 2012. The search for medical technologies abroad: the case of medical travel and pediatric oncology treatment in Argentina. *Technology and Innovation*, 14, pp. 25–38.

Whittaker, A. 2008. Pleasure and pain: medical travel in Asia. *Global Public Health*, 3, pp. 271–90.

Whittaker, A. and Speier, A. 2010. 'Cycling overseas': care, commodification, and stratification in cross-border reproductive travel. *Medical Anthropology*, 29, pp. 363–83.

Whittaker, A., Manderson, L. and Cartwright, E. 2010. Patients without borders: understanding medical travel. *Medical Anthropology*, 29, pp. 336–43.

Wood, E., Yip, B., Gataric, N., Montaner, J., O'Shaughnessy, M., Schechter, M. and Hogg, R. 2000. Determinants of geographic mobility among participants in a population-based HIV/AIDS drug treatment program. *Health and Place*, 6, pp. 33–40.

10 Cycles of violence, girlhood and motherhood

Family formation in Guayaquil's shantytowns

Alysa Handelsman

Families in the shantytown communities of Guayaquil, Ecuador, represent a kinship model that is in constant motion, bolstered by cycles of bloodshed and fractured blood ties. To demonstrate the strategic ways in which families and childhoods are both created and dissolved, I frame this chapter with stories from the girls and women I worked with during more than two years of ethnographic fieldwork. In particular, I describe the physical and emotional movement of daughters and mothers, that is the reasons and the ways in which they cycle from one home to another, from one family to another and from one experience to another. These cycles are certainly complex and contradictory, and they are often triggered by the decisions of girls and women as they measure the stakes of enduring versus escaping violence and abuse.

By analysing such violence alongside shantytown cycles of childhood, girlhood and motherhood, this chapter both poses and complicates the question that women constantly asked as they shared their personal stories: why does history repeat itself? Although young girls see their mother's struggles and hopes for them to have different futures, they often become mothers at a young age as well (often on the cusp of teenagehood), move through multiple marriages, endure abuse and ultimately hope for their children's futures to be different from theirs. In her home in one of Guayaquil's northern shantytowns, Lola, nearly 50 years old, introduced me to three of her young grandchildren whom I had never met before. She told me, 'Mire como la historia se repite' (Look at how history repeats itself). She concluded:

> My mother abandoned my father and my brothers and me; my daughter abandoned her son who I have raised and is now 13; and now my son's wife has abandoned him and his three children, and I will raise them, too . . . history always repeats itself.

I shared my own reflection with Lola that day. Like several other mothers I worked with, I saw her actions as interrupting these intergenerational cycles that they all discussed and feared.

Lola thanked me for my reflection, but questioned whether shifting a cycle off course was enough. She questioned the extent of her power over her children and grandchildren's futures, worrying that her husband's wrath and the milieu of violence within which kids grew up could outweigh the love and affection she gave them. Certainly, the experiences of my research collaborators demonstrate the challenges of growing up and of raising and protecting one's family in Guayaquil's shantytowns. Among these challenges, I focus in this chapter on the common cycles of girlhood and young motherhood as well as the strategies collaborators employ to shift these cycles off course. I use my research collaborators' stories to demonstrate the active role girls and women take in their everyday lives – although within the structural and institutional limits – and how their search for safe spaces through new relationships creates families and influences how we conceptualise 'childhood', 'girlhood' and 'motherhood'.[1]

Guayaquil's shantytowns

This chapter is based on my ethnographic fieldwork (September 2012 to October 2014) in Guayaquil, Ecuador's largest city. Located on the country's Pacific coast, Guayaquil has a population of approximately 4 million. It is a city of extreme socio-spatial segregation, modelled and organised by the competing interests, needs and imaginations of the Municipality and of Guayaquileños from the poorest to the wealthiest neighbourhoods (e.g. Andrade, 2005; Garcés, 2004). It is Guayaquil's shantytowns – the city's poorest communities on its southern, northern and eastern sectors – in which my research collaborators live. These neighbour-hoods form a perimeter around Guayaquil. Their physical structure of unpainted grey brick and bamboo is always visible from a distance, surrounding the city in a perimeter of poverty. Nearly half the city lives in these shantytowns. Many of these homes do not yet have basic household services – even running water – and roads remain unpaved, most notably on the north side. The segregation of the city and the unequal distribution of resources and services reflect the structural violence that frames people's experiences in Guayaquil, particularly the experiences of the poor. Structural violence is not always clearly visible; it is embedded in the systems – political, legal, economic, social – that influence people's decisions, possibilities and everyday lives (Tyner and Inwood, 2014). Although it is primarily domestic violence that I describe in this chapter, the experiences of my research collaborators cannot be understood without also acknowledging that the structural violence of poverty (which is evident through the city's socio-spatial segregation, for example) impacts their everyday experiences and possibilities.

Residents of all ages from Guayaquil's shantytowns move across the city to work in houses as maids, bodyguards and chauffeurs or to work as lower-level employees in businesses: bag boys at grocery stores, servers at restaurants, custodians at shopping malls, etc. Men also work independently as electricians, plumbers and painters or on city streets and buses. Others sell lottery tickets and bags of limes and mangos, running from car window to car window to make their sales. Women carry babies on their backs as they sell roses and candy. Children are also on the street as they juggle tennis balls, wash windshields and beg for money at traffic intersections. People with visible tumours, in wheelchairs and on crutches move along the intersections during red traffic lights holding signs that describe how their disability makes their employment impossible. These are the images outlining the city.

Although Ecuador's National Institute of Statistics and Census Data (Instituto nacional de estadística y censos, INEC, 2014) indicates that in urban areas such as Guayaquil approximately 17 per cent of the population lives below the poverty line and only approximately 4 per cent lives in extreme poverty, the ways in which they classify 'poverty' do not coincide with how it is lived.[2] The majority of the people with whom I worked earned less than the minimum wage, working informally selling food, cleaning different people's houses and cooking. Their average household income is less than US$500 per month and the average household size is seven.

During my fieldwork I conducted participant observation, individual and group interviews, and also focus groups at a non-profit foundation for street children, in children's homes and neighbourhoods, and with children and their families across the city.[3] Working with the children in their peer groups, by themselves, with their families, with people from other socio-economic classes and in several spaces across the city, I had the opportunity to gain a broad understanding of the social and spatial forces that the children confront and make sense of on a daily basis, especially the ways in which girls make sense of the violence that structures their everyday experiences.

Fifty children between the ages of 8 and 18 actively participated in one or more of the phases of this research project, along with their 25 female guardians. Some of the children had been abandoned by their biological parents and were being raised by aunts, cousins, siblings, grandparents or in group homes. Experiences of emotional, physical and/or sexual abuse are commonplace. These children talk about murder and abuse casually, and they even laugh at times as they narrate details and events that are terrifying and severe (for related ethnographic examples, see Márquez, 1999; Goldstein, 2003). But all these events form part of their everyday lives. For them, danger and violence have become routinised and expected (for a review of danger, violence and childhood, see Korbin, 2003). However, children and their families are not passive in the face of violence and their responses to hostile environments set kinship in motion as they strategically adjust familial bonds for their wellbeing and survival.

Blood ties and bloodshed: the interconnection of violence, childhood and family formation

The socialisation and development of children is structured by their surroundings. For example, the violence girls endure in their shantytown homes, their interactions with their friends and neighbours on the street, and their decisions to move from one home to another are interconnected with their socialisation and development. Although streets are considered 'sites of passage' for street children (Kovats-Bernat, 2006: 16), the shantytowns fulfilled this role for the children I worked with because it is in these communities that they learn, grow and develop skills and interests. In communities like Guayaquil's shantytowns 'violence is a perva-sive part of the social landscape. There is, in essence, no escape, just degrees of involvement' (Wolseth, 2004: 218). Violence certainly frames the home and street life of children growing up in the shantytowns, but the children I worked with do not strive to escape their neighbourhoods; rather, their forms of escape involve creating or joining new families, sometimes on the same neighbourhood block.

During one of my neighbourhood visits on the East Side, a group of women pointed to their surroundings and Irene asked me: 'What kind of future is possible when our children are surrounded by drugs, prostitution, gangs?' When children are raised in neighbourhoods governed and terrorised by drug dealers and gang bangers, and in households in which their mothers are knocked unconscious, or in which their siblings are career criminals, these are the ways of life they know and contemplate as they determine their 'degrees of involvement' – these are their 'sites of passage'.

Anthropologist Lawrence Hirschfeld insists that studying children is crucial to our understanding of the ways in which learning can and does happen because 'anthropology is premised on a process that children do better than almost all others, namely, acquire cultural knowledge' (2002: 624). Acquiring cultural knowledge from these environments, however, does not mean that children necessarily imitate what they learn from their surroundings; in fact, a majority of the children I worked with spent their days in their neighbourhoods playing soccer in empty fields, watching TV, gossiping with friends and playing Bingo with their neighbours. Regardless of the extra-curricular activities they choose to join, the children in these neighbourhoods are influenced by their surroundings and they grow up and mature accordingly. Violence in particular plays an important role in the maturing process. Violence disrupts child/adult boundaries in the Western sense by 'exceeding its limits' and by pushing children 'over the boundaries of what it is to be a child' (Caputo, 2001: 183).[4]

One morning at the non-profit agency, as I pinned a map on the wall for our English lesson, Darcy sat and asked me questions about the US states. As she practiced pronouncing Alabama, Alaska, Arizona and Arkansas, I noticed the red marks on her legs. When I asked her what happened, she said her father had ruined her birthday by beating her with a stick. Later, when I spoke with him during report card pick-up, he explained that he is strict with his daughter because he wants to protect her and punishes her when she tries to leave the house to talk with older boys. In this way, he ensures adult control. He feels like he failed with his oldest child who is already a teenage mother, and he does not want history to repeat itself with his youngest: 'Ya perdí a la otra. No quiero que la misma historia vuelva a suceder'.[5]

The parents I met through my research struggled as they worked to protect their children. In general, they believed that physical punishments are the only way their children will learn to respect orders and make good decisions; beating their children is one way of showing them love. As a child, Elisa said, her mother burned her hands when she stole money from a neighbour; after that, she never stole again. Now, however, she concluded that children's rights have created delinquent youth who do not know how to respect because they are not disciplined.

Through my fieldwork I witnessed the routinisation of danger and fear in people's everyday lives and how frequently the most severe forms of violence are perpetuated within households and by one's own family. The violence described in this chapter focuses on the (threat of) physical and sexual abuse within households and how such abuse shapes childhoods, girlhoods and motherhoods. Abuse motivates family formation, stimulating a desire for change by creating or joining new families through marriage and reproduction. Through a husband, for example, girls can escape their abusive households and find new families through in-laws and through the babies they conceive. In some cases, mothers must decide between the overall wellbeing of their family versus the wellbeing of their daughters as they decide whether or not to take a new husband. A stepfather in Guayaquil's shantytowns simultaneously (and paradoxically) represents a potential abuser and a potential source of income and support. The loss of blood brought about by abuse and by the fractured bonds between children and their biological parents can give way to a cycle of bloodshed and blurred blood ties that traces the movement of family formation and its corresponding phases of girlhood and motherhood.

Biology versus bonds that endure: in-laws

Early anthropological studies of kinship in particular focused on blood as a way to organise relatedness (Carsten, 2011). Biology structured understandings of familial relationships and was used to make evolutionary determinations to compare and render inferior different communities' levels of development and sophistication (Feeley-Harnik, 1999). In the shantytowns, however, it is bloodshed that frames kinship more than biological blood ties. Kinship relationships are adjusted based on strategic decisions of women and children as they determine the stakes of enduring versus escaping abuse. Although biological relationships may at times provide a sense of love and belonging, they are not binding and are often not the ones that shantytown mothers and children deem the best fit for their wellbeing.

Many of the girls I worked with take action against the abuse they endure by searching for a way out. They seek out possibilities for love and belonging from young men who make promises before, during and (sometimes) after sex of the future they can share together: a house, food, money and babies (for a discussion of 'push–pull factors' that street children confront as they choose between the streets and their homes, see Strehl, 2011). Seventeen-year-old Yolanda, for instance, was thrilled to learn that she was going to be a mom. She felt the pregnancy secured her relationship with her husband (Nestor) whose love she had always questioned. She found comfort believing they would be tied together for life through blood, through their baby. By becoming his wife, she moved in with his family and hoped to leave behind her abusive past. During an afternoon walk, her next-door neighbour Lulita told me the rumours that were circulating on their block of the sexual abuse Yolanda had endured as a child from her own father.

Marlene cried when she told me what her daughter-in-law, Yolanda, had almost done – the potential blood loss. At eight months pregnant, Yolanda became hysterical when her husband arrived late from work. She proceeded to pull out a knife and threatened to stab herself, putting an end to her life and her baby's. Lulita heard the screams from next door and ran over in time to witness Nestor struggle to take the knife from his wife's hands. They were both cut in the process. Yolanda recognised their baby was not enough to secure their bond; their relationship was not binding. Not unlike Yolanda, girls often feel disposable upon becoming pregnant or upon giving birth because their allure appears to wear off for their partner. As I have learned from interviews and observations, some girls who are abandoned or cheated on consequently become resentful of their children, and this resentment supports their decision to leave them behind.

Although Yolanda's jealousy and insecurity over her marriage tainted her relationships with her new family and led to the end of her union with Nestor, she still lives in Marlene's home. Now, however, she shares a room with her mother-in-law and her new-born son, and Nestor sleeps separately. Several girls I interviewed and spoke with in the shantytowns, like Yolanda, expressed their happiness and willingness to live with their husbands' family. Not only is the new home a safer space than the one in which they grew up, but they love their mother-in-law. Girls often see their mother-in-law as the mother they never had. In fact, some of the young girls are more excited by their relationships with their mother-in-law and sisters-in-law than with their husband. Angela, for example, was 15 when she went to live with her husband's family; she felt loved and protected by her mother-in-law, especially when her husband entered a rehabilitation facility for drug addiction. Even before her husband was sent away, however, she told me that she was grateful for her husband because, through him, she finally had a mother who truly cared about her.

This relationship helps young girls with their transition into adult- and mother-hood. The high turnover rate of wives in the shantytowns, however, normally implies that these bonds, while important, are also temporary. These are not bonds that endure, but while they last they can provide girls with an environment that allows them to endure (e.g. Weston, 2001). Through their husband, women gain a family of their choosing.

Others, however, are not as lucky in their new homes (for a discussion on domestic violence undertaken by extended family members, see van Vleet, 2002). Morelia, for example, told me that she ran back to her mother's house when her sister-in-law tried to force her into prostitution as a way of contributing to the household. Girls and women can also face criticism from in-laws and encounter violent attacks from their husbands on account of rumours circulated by his family. Typically, these rumours are based on claims of the wife's infidelity. These negative experiences demonstrate what some of the stakes are for these new unions and what decisions girls and women make as they navigate these risks alongside their alternatives. While we sat and watched her son and daughter play at a park across from my house, Mireya described the last time her husband beat her. His sister told him of her supposed affair with a neighbour. She said that she sent all of her children outside and locked the door. She confronted her husband and fought him with everything she had. The final thing she remembers before she blacked out was hoping that her children would survive without her.

Blood flow: the menstrual cycle

Menstruation is another cycle that guides family formation in the shantytowns. Menstruation symbolises a girl's entrance into womanhood and the possibility of reproduction. Blood's 'flow within and from the body is closely bound up with life itself' (Carsten, 2011: 29), and by being bound up with life the menstrual cycle's potential to expand one's family can be a source of anxiety in the shantytowns. I interpret young girls' menstrual blood flow as taking one of two directions: one is forced and the other is desired.

For girls who are already enduring sexual abuse, their menstrual cycle gives this abuse a new level of risk because they may become pregnant. A DNA test uncovered the truth behind one family's blood ties in Guayaquil's south side. Carmen, a 13-year-old, recently discovered that her step-grandfather is actually her biological father. Along with her grandmother, this is the man who raised her and treated her like a daughter. I asked her grandmother if she would stay with her husband. She told me that her husband would have to ask God for forgiveness. God will judge him, she told me – not her. Carmen's mother was Carmen's age when she gave birth and decided to flee her house and abandon her daughter. Thirteen years later, her story has been revealed and Carmen is left to make sense of the relationships that have structured her life thus far.

For other girls, their cycle represents their interest in exploring and starting their own sex lives, and this new phase often takes the shape of a baby. During lunch one afternoon, Norma cried as she told me that her daughter had started her first menstrual cycle. She was uncertain that she could protect her or avoid losing her to sex. Sex claimed her other daughter at 13 years old. Now that Morelia, her oldest, is 16 with two daughters of her own, Norma has forced her to receive birth control injections that prevent pregnancy. Norma is actively intervening in her daughter's menstrual cycle, regulating her blood flow. Her son-in-law became angry by this intrusion, exclaiming that his wife would finally mature and become responsible only after having her third child.

As girls actively attempt to create new families for themselves, their decision to have children is a way of securing the marriage and a way of inspiring confidence in a husband, convincing him not to doubt her and her fidelity. Carito was 15 years old when she moved in with her husband; she was scared because she was a virgin. She waited two weeks before having sex with him. From the beginning, she said, 'I told him I didn't want to become pregnant right away, because I was still young' – 'iba a ser una niña criando a un niño' [I would be a child raising a child]. Carito decided to stop taking her birth control pills after her doctor told her she could become sterile. She didn't know she was pregnant until the middle of her second trimester. She said that as soon as she heard her baby's heartbeat, 'ya comencé a cambiar' [I started to change]. She doesn't feel like an adult, she explained, but now that she's going to be a mom, she said, she knows she's no longer a child.

Sex for Lourdes (a 15-year-old) was her opportunity for motherhood – she was thrilled about her pregnancy, exclaiming that she had waited her whole life to be a mother; her own mother had abandoned her and her siblings, and she had never forgiven her or understood how a mother could choose to deny her children her love. At 15, Lourdes moved in with her mother-in-law while her husband lived in the police quarters for his training. She said that she suspected he had already been unfaithful, but, she told me, as long as he loved his baby and gave them everything they could want or need, she did not care how many other women he was with. She said she would be happy as long as she had her daughter. She told me in an interview before giving birth that once her baby was born: 'for the first time in my life, I will not be alone'.

Lourdes succeeded in attaining a safe space and new relationships through her mother-in-law, her husband and her daughter. Simultaneously a girl and a mother, Lourdes made decisions to create a better life – a better childhood – for herself through motherhood; she would no longer experience it alone – and she would be able to provide her daughter with a mother's love, which she never had. She was at once recreating the cycle by becoming a teenage mom and aspiring to re-route it by choosing to be a different type of mother.

The overlap of girlhood and motherhood in the case of Lourdes remains evident with older mothers like Jaqui who is 32 years old. As one of the coordinators at a non-profit organisation, I organised a fieldtrip to a historic park in Guayaquil. When I called to confirm her son's attendance, Jaqui informed me that she would also be joining us. She told me that she never got to be a kid and go on fieldtrips when she was her son's age; it was therefore only fair that we invite her as well. This example and many others from my fieldwork demonstrate that by giving their children a childhood they never had, mothers also seek to experience it simultaneously, reclaiming their own childhood through motherhood.

Because many of the girls I worked with already have children and became pregnant as young as 12 years old, I also asked them directly if they thought that a 12-year-old could be a mother and a child simultaneously. My young collaborators disagreed on this point on first sight. Tomy, for example, insisted that as soon as you become a mother your childhood ends. Saruka, however, said that when your mom helps you raise your baby, you can still be a child because you do not have as much responsibility. Kaín interjected by adding that 'not everyone feels grown up at the same age'. Mayra elaborated on this aspect further when she said that as one grows up the body changes, but those changes don't necessarily mean that one is emotionally mature. Although the children shared diverse perspectives, their responses highlighted that physical and mental maturity do not necessarily go hand-in-hand. The boundaries of childhood, girlhood and motherhood are fluid because the speed and trajectory of these are driven by the children's relationships with people and spaces – specifically relationships from their past, present and potential ones they envision for their future.

Sex, violence and vulnerability: sacrificing girls for family wellbeing

Dwellings in the shantytowns are small and many children are raised in a space of 25–50 square metres in which an entire family sleeps in the same room. From birth, children can therefore witness their parents or guardians having (violent) sex. Sex forms part of their everyday lives, part of their sites of passage. At the non-profit, a psychologist shared several cases with me. The first was of two sisters and a brother who described the different men their mother would allow into their home and the sounds and screams she would make. A sheet formed the dividing wall between her bedroom and her children's. During another session, a 10-year-old girl asked why her stepfather squirted milk from his body. The question the psychologist was never able to answer was why the young girl knew this: was she forced to watch or was it the inevitable eavesdropping of living in a one-room house?

Even married women are subject to rape – sex is forced upon them by husbands they no longer love. As several women told me during interviews and informal conversations, economic dependency – exacerbated by a growing number of children – forces them to maintain their marriage and endure abuse and assault for the sake of their family. Norma worked as a live-in maid during her teenage years and was sexually abused by her employer. Although she never loved her husband, he represented a way out and she ran off with him. She is determined for her daughters not to endure her same fate. She and other mothers often fear leaving their children home alone because of the dangers that overwhelm their neighbour-hoods. Sometimes, however, they become even more distressed if their children are with stepfathers or other male relatives. Norma, for example, built a separate house on her property for her three sons who are addicted to drugs. She and her youngest daughter sleep in the original house. One day, Elsita told Norma that she was scared because her brother snuck inside her *toldo* (mosquito netting) the previous night. Elsita explained that just before he fell asleep right next to her, he stared down at her with a panicked look in his eyes. Norma told me that her sons love their sister, but if they are high on drugs they might not realise she is their sister and, unknowingly, abuse her sexually. Next time that happens, she instructed Elsita to scream.

The potential for sexual abuse within her household is also a concern of Rosario's. Nevertheless, she told me she will fight for her husband until the very end, threatening any woman who tries to steal his attention. He might not be the most handsome man or the most interesting or talkative one, but he is a hard worker and he is responsible, she explained. He looks out for all her children, even the older ones who are not his. He gives them money for food and school supplies, and he undertakes necessary home repairs. When I asked her about her plan to build an indoor bathroom for her daughters' room, she explained that her husband is a good man, but that it is only natural for him to fall into temptation. It is safer for everyone, she said, if the girls have their own bathroom with a lock and a door.

Not unlike the separate house Norma built, Rosario hopes that a bathroom will protect her daughters. She is looking out for her children's wellbeing by fighting for a husband who provides for them. It is rare to find a hardworking and responsible man, she emphasised. In forming her family, Rosario decides that his positive attributes outweigh the risk of the girls' abuse, of their bloodshed. The potential sexual temptation that could get the better of him is an inevitable risk, and a risk worth taking.

Based on my research, I have concluded that girls who are born in the shantytowns will likely endure sexual abuse at some point in their lifetime, making their positionality exceedingly vulnerable. I was taken aback during interviews in which young girls spoke about being sexually molested and attempted to lessen the severity of this abuse by explaining that they were 'only' touched and that it could have been worse.

This section underscored a mother's challenges because the decisions they make for the wellbeing of their families necessarily involves risk and uncertainty. These examples complicate our notions of wellbeing by demonstrating the sacrifices of daughters and mothers for the sake of their families: just as daughters may be sacrificed, mothers sacrifice themselves and their bodies so that their husbands continue to provide for them. Sexual abuse in many ways, therefore, became normalised and a strategy to ensure a family's wellbeing. Although the prevalence and brutality of the physical and structural violence presented in these narratives is unquestionably extreme, they align with the sexual politics experienced and employed in other contexts (e.g. Cole, 2010). Being well is more than feeling and being physically and mentally well on the individual and subjective level; it is tied to the social, physical and economic wellbeing of the family unit in its entirety. The physical and mental wellbeing of individual family members, especially girls, becomes entangled within the wider wellbeing of the family they were born into or choose to live with.

Conclusion

In many ways, this chapter goes against the common understanding and boundaries of 'wellbeing' and 'childhood'. First, cycles of girlhood and motherhood overlap and complicate girls' and children's notions of what childhood or wellbeing is and what it should be: the contradictions between the normative concept of 'childhood' and their own childhood.

This study has also shown that diverse pathways and cycles of childhood/girlhood exist. Girls and women cycle through their changing roles and relationships as the overlaps, the continuity and the breaks begin to demonstrate the risk, despair and cautious optimism that go into their movement as well as their attempts to stand still. A majority of the pregnancies were planned and, for some, were lined with the hope of giving their children an upbringing and a life they always longed for themselves. For others, it was a way of securing their relationship with their partner; and for many of the girls it was a way to move to a new home with a new family – a safe(r) space.

Although the actions of teenage mothers described are not unique to Guayaquil's shantytowns, my research challenges common notions of teenage pregnancies being 'accidental', 'naive' and 'ignorant'; rather, for the young mothers in this study, pregnancy was often an intentional strategy to actively make changes in their lives – to enhance their subjective wellbeing. Sometimes the changes they sought for safe spaces and new homes were connected to the abuse they faced in their own homes; other times new relationships were sought as a means to feel love, a sense of belonging and to attain physical and/or emotional places of well-being. These young mothers were active agents in designing their wellbeing within the structural constraints they faced.

Second, violence – structural and physical – as described in this chapter complicated not only the collaborators' framing of wellbeing, but also their experiences and notions of childhood 'by exceeding its limits'. Violence disrupts child/adult boundaries because it pushes 'children over the boundaries of what it is to be a child' (Caputo, 2001: 183). These 'limits' reflect an implicit ideology of a child as a young person in need of protection. However, as this chapter demonstrates there are no clear boundaries between 'childhood' and 'adulthood'; no clear boundaries of who is in need of protection; rather, forms of protection are often unconventional and contradict common sense understandings for fostering wellbeing of vulnerable groups.

Structural poverty and violence shape the shantytown's 'sites of passage' and create uncertain everyday living conditions – children do not always know if there will be enough food to eat, or what forms of abuse they may encounter, or whether or not their wellbeing may be sacrificed for what is considered a 'greater good' by others. Delaney (2001), for example, references *Genesis* and the story of Abraham to argue that kinship is founded upon sacrifices and violence – or 'the greater good'. As shown in this chapter, Guayaquil's shantytowns are no exception. In the case of Carmen, for example, her life began when her mother was raped. Similarly, Norma started her family as a means of escaping sexual abuse. The ultimate violent act, Delaney proposes, is the sacrifice of one's own child. Rosario chooses to expand her family and to bind them to a stepfather, potentially sacrificing her 11- and 13-year-old girls for the sake of her nine children. A private bathroom is her attempt to delay the sacrifice. In contrast, Yolanda's jealousy led her to take the knife in her own hands and perform violence upon herself and, in the process, upon her unborn child. She imagined that through this sacrifice – or the threat of sacrifice – her husband would love her. Other girls who say they have waited their whole lives for motherhood are ready to terminate their relationships with their children as quickly as Yolanda. Their frustration and boredom with motherhood, coupled with their husband's growing disinterest in them and their child, motivates them to sacrifice their children by leaving them behind and reclaiming the rest of their youth.

To summarise, the childhoods I sketched in this chapter were traumatic and prematurely adult in many ways. Their stories frequently disrupted the normative notions of childhood. Children themselves highlighted this complexity by referring to themselves as 'niños' (children) while concurrently speaking of the childhood they never had. As they grew older, many attempted to recreate the innocent childhoods unknown to them through participating in activities associated with childhood (e.g. fieldtrips). By acknowledging and addressing the complexities of seeking wellbeing in violent childhoods (e.g. through teenage pregnancies), we are in a unique position to better identify appropriate resources for children's wellbeing as they grow and age. We have to acknowledge the fluid and unorthodox boundaries of *childhood* wellbeing.

Notes

1 The experiences and cycles described in this chapter are common across poor communities in Latin America and the world. Space limitations, however, do not allow for a complete discussion of the structural violence of poverty and the ways it manifests itself in Guayaquil, but I do aim to stress that my collaborators are active agents in confronting and transforming their everyday living environments (see also Scheper-Hughes, 2004).
2 The INEC classifies poverty as a monthly income of US$83.29 and extreme poverty as US$46.94. The minimum wage for 2015 was set at US$354 per month, and the 'Canasta Familiar Básica' is set at US$668.57. This 'Basic Family Basket' represents the products that families need per month to survive. These baskets are based on a four-person household and are comprised of 75 products from a list of 359, including food, beverage, household expenses, clothing and miscellaneous goods.
3 Spanish is my first language. My mother and a majority of my family are from Guayaquil I frequently travelled to and lived in the city for an extended period of time. Communication and translation during my fieldwork were never problematic for me or for the people with whom I worked.
4 In this chapter, I use English equivalents of the ways my collaborators referred to themselves, each other and their everyday experiences: 'niños' (children), 'niñez' (childhood), 'adolescentes' (adolescents/teens), 'jóvenes' (young/youth), 'ser madre' (to be a mother/motherhood), etc. These labels coincide with the UN definitions of 'children' (18 years and under) and 'youth' (14–18 years). Ecuador's constitutional code for children and adolescents – El Código de la Niñez y Adolescencia – defines a child [niño (boy) / niña (girl)] as a person who has not yet turned 12 and an adolescent as being between 12 and 18. There is an additional clause that states that if there are any doubts about a person's age, one must assume the person is a child before assuming she is an adolescent. The person must also assume she is an adolescent before assuming she is over 18 years old. I use 'young people' and 'children' interchangeably to refer to the groups I have worked with (18 years and under) to reflect my collaborators' categorisation.
5 'I already lost the other one. I don't want the same thing [literal translation: the same history] to happen again'.

References

Andrade, X. 2005. 'Más Ciudad', menos ciudadanía: renovación urbana y aniquilación del espacio público en Guayaquil. In: F. Carrión and L. Hanley, eds. *Regeneración y Revitalización Urbana en las Américas*, pp. 147–68.

Caputo, V. 2001. Telling stories from the field: children and the politics of ethnographic representation. *Anthropologica*, XLIII, pp. 179–89.

Carsten, J. 2011. Substance and relationality: blood in contexts. *Annual Review of Anthropology*, 40, pp. 19–35.

Cole, J. 2010. *Sex and salvation: imagining the future in Madagascar*. Chicago, IL: University of Chicago Press.

Delaney, C. 2001. Cutting the ties that bind: the sacrifice of Abraham and Patriarchal kinship. In: S. Franklin, and S. McKinnon, eds. 2001. *Relative values: reconfiguring kinship studies*. Durham, NC: Duke University Press, pp. 445–67.

Feeley-Harnik, G. 1999. 'Communities of Blood': the natural history of kinship in nineteenth-century America. *Comparative Studies in Society and History*, 41(2), pp. 215–62.

Garcés, C. 2004. Exclusión constitutiva: las organizaciones pantalla y lo anti-social en la renovación urbana de Guayaquil. *íCONOS: Revista de Ciencias Sociales*, 20, pp. 53–63.

Goldstein, D.M. 2003. *Laughter out of place: race, class, violence, and sexuality in a Rio Shantytown*. Berkeley, CA: University of California Press.

Hirschfeld, L. 2002. Why don't anthropologists like children? *American Anthropologist*, 104, pp. 611–27.

Instituto nacional de estadística y censos (INEC). 2014. *Compendio Estadístico 2014*, Government of Ecuador. Available at www.ecuadorencifras.gob.ec/documentos/web-inec/Bibliotecas/Compendio/Compendio-2014/COMPENDIO_ESTADISTICO_2014.pdf (accessed 20 February 2015).

Korbin, J.E. 2003. Children, childhoods, and violence. *Annual Review of Anthropology*, 32, pp. 431–46.

Kovats-Bernat, J.C. 2006. *Sleeping rough in Port-au-Prince: an ethnography of street children and violence in Haiti*. Gainseville, FL: University Press of Florida.

Márquez, P.C. 1999. *The street is my home: youth and violence in Caracas*. Stanford, CA: Stanford University Press.

Scheper-Hughes, N. 2004. Dangerous and endangered youth: social structures and determinants of violence. *Annals New York Academy of Sciences*, 1036, pp. 13–46.

Strehl, T. 2011. The risks of becoming a street child: working children on the streets of Lima and Cusco. In: G.K. Lieten, ed. *Hazardous Child Labour in Latin America*. Leiden: International Research on Working Children (IREWOC), pp. 43–65.

Tyner, J. and Inwood, J. 2014. Violence as fetish: geography, Marxism, and dialectics. *Progress in Human Geography*, 38(6), pp. 771–84.

van Vleet, K.E. 2002. The intimacies of power: rethinking violence and affinity in the Bolivian Andes. *American Ethnologist*, 29(3), pp. 567–601.

Weston, K. 2001. Kinship, controversy, and the sharing of substance: the race/class politics of blood transfusion. In: S. Franklin and S. McKinnon, eds. *Relative values: reconfiguring kinship Studies*. Durham, NC: Duke University Press, pp. 147–74.

Wolseth, J.M. 2004. *Taking on violence: gangs, faith, and poverty among youth in a working-class Colonia in Honduras*. PhD thesis, University of Iowa.

Part III
Gardens, greens and nature

11 Urban green spaces and childhood obesity in (sub) tropical Queensland, Australia

Debra Flanders Cushing, Harriot Beazley and Lisa Law

The subtropical and tropical climates of Queensland, Australia are ideally suited to growing fruits, vegetables and herbs throughout much of the year. Farmers' markets and community gardens are increasingly popular in both urban and suburban areas and, for some population groups, contribute to the promotion and availability of fresh produce. However, food networks are complex and the availability of fresh produce for some citizens does not necessarily result in increased availability or behaviour change for all (e.g. Connelly *et al.*, 2011; McClintock, 2013). This chapter is concerned with the availability of fresh produce in light of rising levels of overweight children and childhood obesity in Australia.

Considered to be one of the most serious public health issues of the twenty-first century by the World Health Organization (WHO, 2012a), childhood obesity has reached epidemic proportions. In 2013, the American Medical Association recognised obesity as a disease, although there is still considerable debate on how to classify obesity and whether or not it should be afforded this status (Fitzgerald, 2013). Regardless of the current debate, the situation is worsening, and in 2012 *one-quarter* of Australian children aged 5–17 were overweight or obese (Australian Bureau of Statistics (ABS), 2012). If trends continue, by 2025 it is estimated that over *one-third* of 5–19-year-olds will be overweight or obese (Haby *et al.*, 2012).

Research has identified several key drivers in the global obesity epidemic, including our current food systems, the increased supply of cheap, energy-dense foods and persuasive food marketing (Swinburn *et al.*, 2011). When combined with increased screen time and other sedentary behaviours, rising car dependency, as well as less physical activity in general, a large portion of the world's population faces serious consequences of increasing body size.

Focusing on access to healthy food, this chapter synthesises research on food practices, urban food landscapes and childhood obesity, and considers two examples of community gardens in Queensland. Within this context, we reflect on efforts that could raise awareness of healthy eating practices and begin to provide opportunities for children and families to reduce their risk of becoming overweight and obese. The chapter concludes with recommendations for future research on urban green spaces and obesity.

Childhood obesity situation

Globally, the prevalence of overweight and obese children is highest in upper-middle income populations; however, the highest rate of increase in overweight and obese children is among lower-middle income groups (WHO, 2011: 2). The increase in Australia is on the upper end of the spectrum, with an estimated increase of 1 per cent of all children becoming overweight each year. This is similar to the rate of increase in Canada and parts of Europe, and double that of the US and Brazil (Lobstein *et al.*, 2004: 4).

In Queensland 18 per cent of children aged 5–17 are overweight and 8.5 per cent are considered obese (Queensland Health, 2011). Children within disadvantaged populations are at higher risk, and within disadvantaged areas of Queensland, children are twice as likely to be obese as children in advantaged areas (Queensland Health, 2011). Similarly, a study from the Australian Capital Territory (ACT) found that the rate of overweight and obesity is 4 per cent higher for Indigenous kindergarten children compared with non-Indigenous kindergarten children, a statistically significant difference (Hickie *et al.*, 2013).

Many short- and long-term implications of childhood obesity have been identified. Overweight and obese children experience reductions in their quality of life and are at greater risk of teasing, bullying and social isolation (WHO, 2012a: 13). They are also more likely to have factors linked to cardiovascular disease, such as high cholesterol or high blood pressure, type 2 diabetes and various cancers (WHO, 2012b: 11). This trend will undoubtedly increase the burden on health services globally. In 2005 the total direct cost of obesity and overweight in Australia was estimated at AUD\$21 billion (Colagiuri *et al.*, 2010: 260).

Obesity prevention strategies

WHO recognises there is no single intervention strategy to prevent childhood obesity and solutions can come from three areas: governmental structures, population-wide policies and initiatives, and community-based interventions (WHO, 2012a). More specifically, solutions must consider food marketing, nutritional labelling, food taxes and subsidies, and also policies related to agricultural practices and land use, dietary guidelines and breastfeeding (WHO, 2012a). Evidence suggests that fruit and vegetable consumption may assist in weight management and reduced risks of disease; however, research has shown that previous interventions to increase children's fruit and vegetable consumption have not delivered the desired benefits and have had varying levels of impact that often decrease after the completion of the intervention program (Pomerleau *et al.*, 2005: 23). Additionally, other studies have found that simply increasing fruit and vegetable intake is not enough to reduce participants' weight (Tohill, 2005: 16).

A literature review of obesity prevention programmes during early childhood showed evidence of success for programs with a strong focus on social behavioural theory that were designed to impact skills and competencies (Hasketh and Campbell, 2010). Additionally, an analysis of 16 intervention programs for primary and secondary school-age children found the most effective strategies for increasing fruit and vegetable consumption include specific fruit and vegetable messages, hands-on skill building, active provision of fruit and vegetables at lunch and the involvement of parents, teachers and peers (Pomerleau *et al.*, 2005: 41).

Economics also plays a role in accessing healthy food. A 2004 study to analyse the cost of healthy food in Queensland showed that food prices for healthy food increases as 'remoteness' increases (Harrison *et al.*, 2007). For example, the general cost of healthy food is 29.6 per cent higher in very remote areas of Queensland than in major cities, with the cost of fruit, vegetables and legumes being 20.3 per cent higher. In addition, the study found that basic healthy food items are less likely to be available in 'outer regional', 'remote', and 'very remote' areas (Harrison *et al.*, 2007: 11). These findings demonstrate potential inequities in the availability and cost of healthy food, which could result in barriers to accessing healthy food for all people.

Similarly, research in the US identified price as one of the most important factors in determining what food items are purchased, showing a direct causal link between poverty rates and obesity levels (Drewnowski and Darmon, 2005). Longitudinal research in China found that increases in the price of healthy foods were associated with decreased consumption of those foods and an increase in consumption of substitutes, such as flour as a substitute for rice. In addition, increasing the cost of animal protein products showed decreased overall fat consumption. Although this may have a positive effect on obesity levels, it may also have a negative effect on protein-intake levels for the poor (Guo *et al.*, 1999).

Food practices

The foods that are increasingly being eaten by much of the world's population in the twenty-first century, including in Australia, are processed, energy dense, sweet and animal-based foods, with far less fresh fruit and vegetables being consumed by adults and children than in previous generations (Walton *et al.*, 2010). In Queensland, 60 per cent of children aged 5–17 years consume the recommended number (two) of fruit servings per day, and 29 per cent consume the recommended number (five) of vegetable servings per day (Queensland Health, 2011). As a vulnerable population, children are impacted by available food systems because they often lack nutritional knowledge, are unable to weigh short-term gratification against potential long-term negative outcomes and are readily affected by marketing (Swinburn *et al.*, 2011).

Key barriers to children's healthy diets are availability, convenience, taste preferences, peer pressure and parental/school support (O'Dea, 2003). Children are dependent on their parents and caregivers to provide food that will promote healthy diets, growth and development (Birch, 2006). The sociocultural and socio-economic aspects of certain food practices, such as parental food preferences and parents' views of what is an acceptable diet, also impact on a child's eating habits. Children tend to eat the same foods as the significant adults in their lives, and their food consumption practices are also affected by societal changes in patterns of food consumption and availability. In fact, a family-based approach targeting parents as key agents of change has been shown to be slightly more effective at reducing the percentage of overweight children than programmes targeting only children (Golan and Crow, 2004).

Recent research within children's geographies has revealed the power and significance of food practices within parents' and children's daily lives (Punch *et al.*, 2010). Child feeding practices determine the foods, portion sizes, frequency of eating times and the social context in which eating occurs (Birch, 2006; Punch *et al.*, 2010). Punch *et al.* (2010: 229) also reflect on the social construction of 'food moralities' by stating that 'food is meshed into contested moralities about appropriate behaviour for young people and children which are clearly class-based in character'. For example, when cost is an issue, parents can often only afford to eat cheap, processed, energy-dense, high-fat foods often available at fast-food establishments, which can contribute to obesity in children. Furthermore, Kearns and Barnett (2000: 81) investigated the inclusion of a McDonald's franchise within a New Zealand hospital and found 'ambivalence in the moral geography of fast food consumption'. The study revealed that concerns expressed by adults over the unhealthy fast-food concealed the underlying concerns with healthcare and did not take into account the potential comfort of familiar food offered to children in a hospital setting. In addition, a UK study found that middle-class teenagers are more likely than working-class teenagers to be monitored by their parents to strictly limit their snack or junk food intake, whereas working-class children are given more autonomy in their food consumption (Backett-Milburn *et al.*, 2010).

Knowledge of food and food production processes can directly influence food consumption, with increased knowledge of fruits and vegetables enhancing their consumption by children. For example, studies of school gardening programmes in Brisbane suggest that simply having a vegetable garden at school increases student exposure to a wide variety of fruits and vegetables, which can lead to healthier food choices (Morris and Zidenberg-Cherr, 2002; Somerset *et al.*, 2005). Although rural families reportedly have some understanding of 'food moralities' that reflect local environmental and cultural practices, such perceptions regarding the positive benefits of organic and unadulterated food are less evident in the narratives of children and parents living in urban settings (Curtis *et al.*, 2010). Urban families have less direct access to the food production system in a general sense, and backyard sizes are also shrinking (Hall, 2010), thus depriving children of opportunities to grow food or to have unrestricted access to environmental learning (Malone, 2007).

When considering food practices, it is important to understand the obesogenic attributes of environments in which children live (Walton *et al.*, 2010; Swinburn *et al.*, 2011) that promote high-energy intake and sedentary behaviours. Physical, sociocultural, economic and political factors can strongly influence which foods are produced within a society and how available they are, especially to children. In Australia, the food for sale in most public places is characterised as cheap, palatable, high-energy food, which is low in nutrient density (Birch, 2006).

Ironically, however, Queensland is ideally situated to provide fresh, healthy produce. In the next section, we discuss the tropical and subtropical Queensland context for growing and selling healthy produce.

Food landscapes

Queensland is a large state (1,852,642 square kilometres, approximately seven times the land area of the UK, and over three times the land area of Texas), with a wide range of environmental conditions and climate classification groups (Bureau of Meteorology, 2014). Much of the productive land lies within the tropical, subtropical and grassland area, making it prime for year-round food production (Daff.qld.gov.au, 2014). Queensland is known for its prolific production of fruit, vegetables and nuts. It is the largest producer of vegetables in Australia and the second largest producer of fruit (Daff.qld.gov.au, 2014), producing, for example, approximately 93 per cent of Australia's banana crop in 2011–2012 (Business Queensland, 2013).

Community garden programs and farmers' markets

In Australia, as in other countries, farmers' markets and community gardens are being developed and promoted as places to purchase local, fresh, seasonal food of high quality (Joseph *et al.*, 2013). A review of the nutritional implications of farmers' markets and community gardens show that although data is limited, the evidence that does exist suggests positive benefits of these types of community entities, especially in low-income areas (McMormack *et al.*, 2010).

In Queensland, farmers' markets have in recent years become a popular venue for buying fruits and vegetables grown by local farmers. As of 2014, 11 farmers' markets were registered with the National Farmers' Market Association in Queensland and 181 markets were registered throughout Australia (Farmersmarkets.org.au, 2014). In addition, an online resource for local markets (Marketguide.com.au, 2014) lists another nine farmers' markets in Queensland, with a number of variety markets that may also sell produce.

Similarly, community gardens are fast becoming a mainstay in the urban and suburban landscape. The Brisbane City Council lists 39 community gardens, scattered within a 15-kilometre radius of the Brisbane Central Business District (Brisbane City Council, 2014). This development is promising if community gardens can promote healthy eating for residents involved with them. Research on the effectiveness of garden-based nutrition education programs in the US indicates promising impacts related to fruit and vegetable intake, willingness to taste fruits and vegetables, and increased preferences for fruits and vegetables (Robinson-O'Brien *et al.*, 2009). Another US study analysing community garden impacts found that adults in households that included participants of community gardens consumed fruits and vegetables 1.4 more times per day than those households without an adult participant (Alaimo *et al.*, 2008). However, the current evidence is limited and at times contradictory.

How urban green spaces impact childhood obesity

A growing body of international research aims to make links between children, access to green space and obesity (see, for example, Kipke *et al.*, 2007; Nielsen and Hansen, 2007; Dunton *et al.*, 2009; Coombes *et al.*, 2010; Lachowycz and Jones, 2011; Wolch *et al.*, 2011). In general, research suggests that green space use by community members provides benefits in terms of physical and mental health, as well as having follow-on socio-economic benefits. Encouraging walking, providing access to spaces that are psychologically rejuvenating and providing a context to bolster social capital are amongst the benefits of green space (Hartig *et al.*, 1991; Seeland *et al.*, 2009). As Lee and Maheswaran (2011) note, however, establishing a causal link between green space and positive health outcomes is not a simple process.

Studies of green spaces, where age is used as a variable, show there is a decline in green space activity for teenagers and young adults (Ball *et al.*, 2007). This could be related to the appropriateness of the green space available to this group, with many spaces given over to younger children. Other studies suggest that children with better access to green spaces (such as parks and recreational resources) are less likely to have increasing body mass index (BMI) (Wolch *et al.*, 2011).

In conversations about green space with a focus on public gardens there is some evidence to suggest that participation in community gardens has the potential to increase fruit and vegetable intake (Alaimo *et al.*, 2008), even in low-income families (Castro *et al.*, 2013). These green spaces simultaneously engage young learners, build their confidence and social skills, and make connections between themselves and their communities (Block *et al.*, 2011).

School-based gardens have shown particular success, although the criteria to evaluate success are often difficult to measure (Nowak *et al.*, 2012; Gibbs *et al.*, 2013). In general, however, children tend to eat more fruits and vegetables as well as develop a willingness to try new foods (Heim *et al.*, 2009; Guitart *et al.*, 2014). This overall trend toward healthy eating has the potential to reduce BMIs, increase nutritional intake, increase psychosocial wellbeing and positively impact social and environmental behaviour (Block *et al.*, 2011). It is with this in mind that we turn to our first case study of a community garden in Cairns, tropical north Queensland.

Cairns Flexible Learning Centre Community Garden

Cairns is an Australian regional city with a population in 2013 of just over 150,000. Projected growth is anticipated to expand the city to approximately 250,000 by 2036. Cairns is an ethnically mixed urban centre, with 9.3 per cent of the population identifying as Aboriginal and Torres Strait Islanders, and 11.2 per cent of the population speaking a language other than English at home. The Cairns Flexible Learning Centre (FLC) described here is a state-funded Positive Learning Centre for disadvantaged youth and migrant children in grades five to ten. Together with local Free Range Permaculture (FRP), the Centre began a school garden program in 2008. The garden was originally established as a means to engage disadvantaged youth with behavioural issues (see Chapter 12), but the growing population of Cairns has seen a shift in focus to include migrant youth and children more generally. Both the FLC and the FRP saw the impact of 'providing an alternative approach to working with children who, for various reasons, were struggling with the traditional classroom environment' (Kruze, 2010). Although their agenda was not to curb obesity amongst youth per se, it is a garden with the potential to reach a target group vulnerable to poor eating habits and lifestyles that often contribute to overweight and obesity issues.

The permaculture garden is a tropical garden full of exotic edible plants, many of which students have grown in their home countries of Africa and Asia. Indeed, the garden's success was featured in a television episode of *Costa's Garden Odyssey* (Episode 10, Series 2, aired on SBS 20 May 2010, 8 pm) in which students from Cairns, Burundi and Bhutan were able to express the significance of the garden in their own words. Germaine, from Burundi, claims the garden 'helps us remember . . . because a lot of things here [in the garden] . . . we have in Africa . . . Sometimes [we] take [foods] home and give to our parents'. She describes how the students convene at the garden to cook different ethnic foods, highlighting how 'we like sharing our culture with different people . . . [and] learning from other people's cultures'. Multimedia students Surendra and Om are originally from Bhutan, but were raised in refugee camps in Nepal. They also began taking garden vegetables home to their parents and learning how to cook Bhutanese food. Both students were undertaking multimedia education at the FLC and began compiling recipes cooked at the Centre as part of their studies. These connections between the garden, kitchen and classroom culminated in a FLC cookbook *Recipes from around the world*. Linda Saunders, a teacher at FLC, reflects on the creation of the cookbook, stating how it was effective, 'just to see their faces light up and show me a plant that I didn't recognise, and say "we grow this in our country . . . and this is how we use it", which was different to the way it's used here'.

The garden therefore grows food and is productive, while at the same time it brings young people together to learn across cultures. Being incorporated into an educational institution helps the garden fulfil these multiple roles where connections are made between food, community and healthy eating. Although the garden's agenda is not specifically about obesity prevention, it has great potential to reach groups in lower socio-economic backgrounds that are often disadvantaged by food distribution systems and access to cheap, energy-dense food. Moreover, encouraging young people to be active in green spaces such as community gardens can help new migrants, who may otherwise feel excluded in other green spaces in the city, to settle in after they arrive in tropical Australia. In addition, being involved through community gardens may help longer term residents connect to opportunities for healthy living. In this way, the garden encourages young people to be active, eat healthy food and value community and experiential learning.

Northey Street City Farm

The second case study focuses on the Northey Street City Farm in the heart of Brisbane, Queensland's subtropical capital city of 2.2 million people. Northey Street City Farm is a permaculture garden located in the inner city suburb of Windsor, a multicultural community that includes migrants from India, China, Vietnam, Philippines, Malaysia and Fiji (Australian Bureau of Statistics, 2013). As with the Cairns Flexible Learning Centre, the Northey Street City Farm also focuses on experiential learning, encouraging children and young people to grow and eat healthy food and to participate in a variety of community-based activities. The farm hosts community workshops, an edible landscape nursery, a weekly playgroup, school tours and other educational activities presented as a 'living classroom' (Northey Street City Farm, 2014). It provides children with an opportunity to learn about fruits, herbs, vegetables, composting, worms, recycling, chickens, cob-oven cooking and sustainable food systems (as opposed to chemical or genetically modified monoculture systems). The farm encourages community participation, hosting an extremely popular farmers' market on Sunday mornings that aims to provide organic produce direct from the farmers. There is also a children's program on the same day with clay and arts activities to encourage the community to bring their children with them to the market and farm.

The City Farm is an attractive unstructured outdoor environment – a contrast to the structured spaces in which children spend much of their time in the inner-city. For this reason, the daily playgroup is growing rapidly as parents are keen to expose their children to spontaneous outdoor play in such a space rather than the more formal play centres on offer in the city. Similarly, the school education programs offered through the City Farm are extremely popular, introducing children to local fruit and vegetable production, composting and worm farming, as well as the opportunity to pick their own ingredients for homemade pizzas, baked onsite in the cob-oven. Although, as with the previous case study, there is no direct focus on combatting obesity or the benefits of healthy eating within the organised programs, the education staff believe that it is an 'inevitable outcome' that children begin to make healthier choices once they are exposed to the environment.

Exercise is also an important aspect of preventing childhood obesity and encouraging young people to be physically active within community gardens increases opportunities for healthy living. The farm has attempted to establish a partnership with a local school to involve children in the physical aspect of gardening, although, as many schools now run their own gardens, this has not yet been possible. They also have allotments for families to maintain together, which is great exercise for children of all ages.

In June 2014 the community garden was blighted with a soil contamination incident that resulted in the Farm being closed down by Brisbane City Council and being listed on the State Environmental Management register (Stephens, 2014). The findings of the soil testing indicated a number of areas were contaminated by bonded asbestos, but no airborne asbestos was detected. Lead was also detected on the site, although the levels of lead exposure were assessed as 'quite low' by Queensland Health (Stephens, 2014). The Queensland Department of Environment and Heritage Protection (DEHP) also assessed the soil due to the levels of arsenic, lead and chromium that were originally found on the site. It transpires, however, that arsenic and chromium were only found in ash from a fire, and the ash was removed and the area turfed. No other contamination was found in the 32 soil samples tested following the initial discovery.

The State Environment Department allowed operations to continue at Northey Street City Farm once a specific soil contamination management plan was in place to ensure contact with the contaminants is prevented (Stephens, 2014). Under a plan negotiated with the Brisbane City Council, Farm staff addressed the situation by converting the garden to raised garden beds filled with clean soil and manure, and to cover all walkways and bare soil with grass, paving and mulch. The staff members have taken the opportunity to be innovative with their raised beds, converting old bathtubs and pallets into garden beds. They have also transformed their kitchen garden to demonstrate creative ideas for small space gardening, with imaginative examples of vertical gardens and container gardens; ideal for inner-city residents and which may also capture a young person's imagination.

Conclusion

Subtropical and tropical Queensland are uniquely positioned to provide year-round healthy produce to its residents. In addition, community gardens, farmers' markets and community green spaces are becoming more common and evidence indicates these spaces provide significant benefits towards addressing and preventing childhood obesity. However, the rate of overweight and obesity in Australia is on the rise, as it is around the world, indicating a disconnection between the environment and children's food practices. This disconnection suggests a prime area for additional research and program initiatives to build stronger connections between local food production, consumption and obesity prevention.

Although any solution needs to be multi-faceted, key findings from previous research determined that successful initiatives often focus on impacting the skills and competencies of participants (Hasketh and Campbell, 2010). Similarly, our brief case studies exemplify specific efforts to actively engage young people in growing and consuming healthy foods within community settings. Although neither case site is specifically focused on obesity prevention, both work to embed healthy practices into children's daily lives and give them skills needed to change behaviour. As discussed by Kraftl (2013), community settings that are focused on growing healthy food become spaces for alternative education because they enable young people to internalise the process and immerse themselves in a healthy environment. However, city spaces, such as community gardens and farmers' markets, are not without their challenges. Our case studies found that funding for maintenance needs, environmental issues such as soil contamination and lack of participation by community members are just a few of the issues that need to be addressed during planning and implementation. Both examples, however, show that once challenges are addressed, children are able to gain healthy experiences that foster a shift in their eating practices. In addition, enabling diverse young people to grow and prepare their own healthy food could help address some issues regarding cost and availability, especially in regional areas.

Recent food policy explorations have identified several mechanisms that can lead to changes in behaviour, including the provision of an environment for healthy food preference learning and overcoming barriers to the expression of healthy food preferences (Hawkes *et al.*, 2015). Many of the obesity-related policies around the world have focused on economic incentives, such as taxes, and food subsidies, as well as on mandatory nutritional labelling to educate people about the nutritional value of the food they consume. Although schools are a primary setting for implementing healthy eating programs, community gardens and farmer's markets can also accomplish these mechanisms because they relate to food literacy and skills, as well as providing avenues for participation by multiple population groups (Hawkes *et al.*, 2015: 8).

A greater understanding of the direct benefits of community gardens and farmers' markets on food consumption is needed in order to inform policy at the local and national levels. In addition, many of the extant studies are based within the US context. Queensland's unique environmental conditions and year-round growing season requires additional research to address the tropical and subtropical environments. Making important connections between healthy eating and obesity prevention and continuing this type of work could show benefits for the obesity issues society faces.

References

Alaimo, K., Packnett, E., Miles, R. and Kruger, D. 2008. Fruit and vegetable intake among urban community gardeners. *Journal of Nutrition Education and Behavior*, 40(2), pp. 94–101.

Australian Bureau of Statistics (ABS). 2012. *Australian Health Survey: First Results, 2011–12*. Available at www.abs.gov.au/ausstats/abs@.nsf/Previousproducts/A9DF782F15FB2A18CA257AA30014B66E?opendocument (accessed 18 July 2014).

Australian Bureau of Statistics (ABS). 2013. *2011 Census Community Profiles*. Available at www.censusdata.abs.gov.au/census_services/getproduct/census/2011/communityprofile/305031623?opendocument&navpos=220 (accessed 18 July 2014).

Backett-Milburn, K., Wills, W., Roberts, M. and Lawton, J. 2010. Food and family practices: teenagers, eating and domestic life in differing socio-economic circumstances. *Children's Geographies*, 8(3), pp. 303–31.

Ball, K., Timperio, A., Salmon, J., Giles-Corti, B., Roberts, R. and Crawford, D. 2007. Personal, social and environmental determinants of educational inequalities in walking: a multi-level study. *Journal of Epidemiological Community Health*, 61, pp. 108–14.

Birch, L. 2006. Childhood feeding practices and the etiology of obesity. *Obesity*, 14(3), pp. 343–4.

Block, K., Gibbs, L., Staiger, P.K., Gold, L., Johnson, B., Macfarlane, S., Long, C. and Townsend, M. 2011. Growing community: the impact of the Stephanie Alexander Kitchen Garden Program on the social and learning environment in primary schools. *Health Education Behaviour*, 39, pp. 419–32.

Brisbane City Council. 2014. *Find your local community garden.* Available at https://www. brisbane.qld.gov.au/environment-waste/be-clean-green-brisbane/community-groups/ community-gardens-city-farms/find-your-local-community-garden#/?i=1 (accessed 28 July 2014).

Bureau of Meteorology. 2014. *Indigenous weather knowledge – the key climate groups.* Available at www.bom.gov.au/iwk/climate_zones/map_1.shtml (accessed 18 July 2014).

Business Queensland. 2013. *Banana Industry Market Updates.* Available at https://www. business.qld.gov.au/industries/farms-fishing-forestry/agriculture/crop-growing/ banana-industry/updates (accessed 18 July 2013).

Castro, D., Samuels, M. and Harman, A.E. 2013. Growing healthy kids: a community garden-based obesity prevention program. *American Journal of Preventive Medicine*, 44(3S3), pp. S193–9.

Colagiuri S., Lee, C., Colagiuri, R., Magliano, D., Shaw, J., Zimmet, P. and Caterson, I. 2010. The cost of overweight and obesity in Australia. *Medical Journal Australia*, 192, pp. 260–4.

Connelly, S., Markey, S., and Roseland, M. 2011. Bridging sustainability and the social economy: achieving community transformation through local food initiatives. *Critical Social Policy*, 31(2), pp. 308–24.

Coombes, E., Jones, A. and Hillsdon, M. 2010. The relationship of physical activity and overweight to objectively measured green space accessibility and use. *Social Science and Medicine* 70, pp. 816–22.

Curtis, P., James, A. and Ellis, K. 2010. Children's snacking, children's food: food moralities and family life. *Children's Geographies*, 8(3), pp. 291–302.

Daff.qld.gov.au, 2014. *Queensland Agricultural Land Audit.* Available at https://www.daf. qld.gov.au/__data/assets/pdf_file/0006/77505/QALA-Ch03-Statewide.pdf (accessed 18 July 2014).

Drewnowski, A. and Darmon, N. 2005. The economics of obesity: dietary energy density and energy cost. *American Journal of Clinical Nutrition*, 82, pp. S265–73.

Dunton, G.F., Kaplan, J., Wolch, J., Jerrett, M. and Reynolds, K.D. 2009. Physical environmental correlates of childhood obesity: a systematic review. *Obesity Reviews*, 10, pp. 393–402.

Farmersmarkets.org.au, 2014. *Markets Directory | Australian Farmers' Markets Association.* Available at http://farmersmarkets.org.au/find-a-market/ (accessed 18 July 2014).

Fitzgerald, K. 2013. Obesity is now a disease, American Medical Association decides, *Medical News Today*, 17 August.

Gibbs, L., Staiger, P., Johnson, B., Block, K., Macfarlane, S., Gold, L., Kulas, J., Townsend, M., Long, C. and Ukoumunne, O. 2013. Expanding children's food experiences: the impact of a school based kitchen garden program. *Journal of Nutrition Education and Behavior*, 45(2), pp. 137–46.

Golan, M. and Crow, S. 2004. Targeting parents exclusively in the treatment of childhood obesity: long-term results. *Obesity Research*, 12(2), pp. 357–61.

Guitart, D.A., Pickering, C.M. and Byrne, J.A. 2014. Color me healthy: food diversity in school community gardens in two rapidly urbanizing Australian cities. *Health and Place*, 26, pp. 110–17.

Guo, X., Popkin, B., Mroz, T. and Zhai, F. 1999. Food price policy can favorably alter macronutrient intake in China. *Journal of Nutrition*, 129, pp. 994–1001.

Haby, M.M., Markwick, A., Peeters, A., Shaw, J. and Vos, T. 2012. Future predictions of body mass index and overweight prevalence in Australia, 2005–2025. *Health Promotion International*, 27(2), pp. 250–60.

Hall, A.C. 2010. *The life and death of the Australian backyard.* Collingwood, VIC: CSIRO Publishing.

Harrison, M., Coyne, T., Lee, A., Leonard, D., Lowson, S., Groos, A. and Ashton, B. 2007. The increasing cost of the basic foods required to promote health in Queensland. *Medical Journal of Australia,* 186(1), pp. 9–14.

Hartig, T., Mang, M. and Evans, G. 1991. Restorative effects of natural environment experiences. *Environment and Behaviour,* 23(1), pp. 3–26.

Hasketh, K. and Campbell, K. 2010. Interventions to prevent obesity in 0–5 year olds: an updated systematic review of the literature. *Obesity,* 18(1), pp. S27–35.

Hawkes, C., Smith, T., Jewell, J., Wardle, J., Hammond, W., Friel, S., Thow, A. and Kain, J. 2015. Smart food policies for obesity prevention. *Lancet,* 385(9985), 2410–21.

Heim, S., Stang, J. and Ireland, M. 2009. A garden pilot project enhances fruit and vegetable consumption among children. *Journal of the American Dietetic Association,* July, pp. 1220–6.

Hickie, M., Douglas, K. and Ciszek, K. 2013. The prevalence of overweight and obesity in Indigenous kindergarten children: a cross sectional population based study. *Australian Family Physician,* 42(7), pp. 497–500.

Joseph, A., Chalmers, L. and Smithers, J. 2013. Contested and congested spaces: exploring authenticity in New Zealand farmer's markets. *New Zealand Geographer,* 69, pp. 52–62.

Kearns, R. and Barnett, R. 2000. 'Happy Meals' in the Starship Enterprise: interpreting a moral geography of health care consumption. *Health & Place,* 6, pp. 81–93.

Kipke, M.D., Iverson, E., Moore, D., Booker, C., Ruelas, V., Peters, A.L. and Kaufman, F. 2007. Food and park environments: neighborhood-level risks for childhood obesity in East Los Angeles. *Journal of Adolescent Health,* 40, pp. 325–33.

Kraftl, P. 2013. *Geographies of alternative education: diverse learning spaces for children and young people.* Bristol: Policy Press.

Kruze, K. 2010. Costa and SBS TV come to Cairns. *The Permaculture Research Institute,* 18 May. Available at http://permaculture.org.au/2010/05/18/costa-sbs-tv-come-to-cairns/ (accessed 5 October 2014).

Lachowycz, K. and Jones, A.P. 2011. Greenspace and obesity: a systematic review of the evidence. *Obesity Reviews,* 12, pp. 183–9.

Lee, A.C.K. and Maheswaran, R. 2011. The health benefits of urban green spaces: a review of the evidence. *Journal of Public Health,* 33(2), pp. 212–22.

Lobstein, T., Baur, L. and Uauy, R. 2004. Obesity in children and young people: a crisis in public health. *Obesity Reviews,* 5(S1), pp. 4–85.

Malone, K. 2007. The bubble-wrap generation: children growing up in walled gardens. *Environmental Education Research,* 13(4), pp. 513–27.

Marketguide.com.au, 2014. *The Market Guide: where shall we go today?* Available at www.marketguide.com.au/qld.htm (accessed 27 February 2017).

McClintock, N. 2013. Radical, reformist, and garden-variety neoliberal: coming to terms with urban agriculture's contradictions. *Local Environment,* 19(2), pp. 147–71.

McMormack L.A., Laska, M.N., Larson, N. and Story, M. 2010. Review of the nutritional implications of farmers' markets and community gardens: a call for evaluation and research efforts. *Journal of the American Dietetic Association,* 110(3), pp. 399–408.

Morris, J.L. and Zidenberg-Cherr, S. 2002. Garden-enhanced nutrition curriculum improves fourth-grade school children's knowledge of nutrition and preferences for some vegetables. *Journal of American Dietetic Association,* 102(1), pp. 91–3.

Nielsen, T.S. and Hansen K.B. 2007. Do green areas affect health? Results from a Danish survey on the use of green areas and health indicators. *Health and Place,* 13, pp. 839–50.

Northey Street City Farm (NSCF). 2014. Permaculture in the heart of Brisbane. Available at www.nscf.org.au/ (accessed 18 July 2014).

Nowak, A.J., Kolouch, G., Schneyer, L. and Roberts, K.H. 2012. Building food literacy and positive relationships with healthy food in children through school gardens. *Childhood Obesity*, 8(4), pp. 392–6.

O'Dea, J.A. 2003. Why do kids eat healthful food? Perceived benefits and barriers to healthful eating and physical activity among children and adolescents. *Journal of the American Dietetic Association*, 103(4), pp. 497–501.

Pomerleau, J., Lock, K., Knai, C., and McKee, M. 2005. *Effectiveness of interventions and programmes promoting fruit and vegetable intake*. Geneva: World Health Organization.

Punch, S., McIntosh, I. and Emond, R. 2010. Children's food practices in families and institutions. *Children's Geographies*, 8(3), pp. 227–32.

Queensland Health. 2011. *Self Reported Health Status 2011 and Reported Child Health Status 2011: Overweight and Obesity, Queensland*. Available at https://www.health.qld. gov.au/__data/assets/pdf_file/0026/443339/srhs11-active.pdf (accessed 27 February 2017).

Robinson-O'Brien, R., Story, M. and Heim, S. 2009. Impact of garden-based youth nutrition intervention programs: a review. *Journal of the American Dietetic Association*, 109, pp. 273–80.

Seeland, K., Dubendorfer, S. and Hansmann, R. 2009. Making friends in Zurich's urban forests and parks: the role of public green space for social inclusion of youths from different cultures. *Forest Policy and Economics*, 11(1), pp. 10–17.

Somerset, S., Ball, R., Flett, M. and Geissman, R. 2005. School-based community gardens: re-establishing healthy relationships with food. *Journal of the HEIA*, 12(2), pp. 25–33.

Stephens, K. 2014. Northey Street soil testing finds arsenic, asbestos and lead. *Brisbane Times*, 9 August. Available at www.brisbanetimes.com.au/queensland/northey-street-soil-testing-finds-arsenic-asbestos-and-lead-20140808-1024b1.html (accessed 10 September 2014).

Swinburn, B., Sacks, G., Hall, K., McPherson, K., Finegood, D., Moodie, M. and Gortmaker, S. 2011. The global obesity pandemic: shaped by global drivers and local environments. *Lancet*, 378(9793), pp. 804–14.

Tohill, B. 2005. *Dietary intake of fruit and vegetables and management of body weight*. Tokyo: World Health Organization.

Walton M., Waiti J., Signal, L. and Thomson, G. 2010. Identifying barriers to promoting healthy nutrition in New Zealand primary schools. *Health Education Journal*, 69, pp. 84–94.

Wolch, J., Jerrett, M., Reynolds, K., McConnell, R., Chang, R., Dahmann, N., Brady, K., Gilliland, F., Su, J.G. and Berhane, K. 2011. Childhood obesity and proximity to urban parks and recreational resources: a longitudinal cohort study. *Health and Place*, 17, pp. 207–14.

World Health Organization (WHO). 2011. *Global status report on noncommunicable diseases 2010*. Available at http://apps.who.int/iris/bitstream/10665/44579/1/9789240686458_eng.pdf (accessed 27 February 2017).

World Health Organization (WHO). 2012a. *Population-based approaches to childhood obesity prevention*. Geneva: WHO. Available at http://apps.who.int/iris/bitstream/10665/80149/1/9789241504782_eng.pdf?ua=1 (accessed 27 February 2017).

World Health Organization (WHO). 2012b. *Prioritising areas for action in the field of population-based prevention of childhood obesity: a set of tools for member states to determine and identify priority areas for action*. Geneva: WHO.

12 Is 'natural' education healthy education?

A comparative analysis of forest-based education and green care spaces in Germany and the UK

Silvia Schäffer and Peter Kraftl

This chapter initiates discussion about the similarities and differences between nature-based education in different national contexts, concentrating on health and wellbeing. It compares three forms of nature-based education[1] and is based on two separate studies that examined Forest Kindergartens (in Germany) and Forest Schools and Care Farms (in the UK). Given the relative dearth of cross-cultural comparative studies of these forms of education (Kraftl, 2013a), it offers insights into the intersections of health/wellbeing and education, principally according to the teachers and practitioners who operate in nature-based education spaces. This introduction provides insight into three forms of 'natural' education: German Forest Kindergartens (FK) and Forest Schools (FS) and Care Farms (CF) in the UK. The second part of the chapter shows links between wellbeing, health and 'natural' education. Finally, social-scientific research about 'natural' education is introduced.

The concept of FK was established in Denmark (1952) and Sweden (1985), where life with nature is associated with quality of life (Knight, 2009). The concept has spread throughout Europe and beyond. In Germany FK are mostly founded independently by parents or kindergarten teachers. They are very popular, with about 1,500 FK estimated by the 'Bundesverband der Natur- und Waldkindergärten in Deutschland' (Federal Association of Nature and Forest Kindergartens in Germany) at the time of writing (http://bvnw.de/uber-uns/). This means that 2.8 per cent of all day care facilities are FK.[2] FK in Germany have existed since 1968, but were only officially recognised as a form of day care in 1993 (Miklitz, 2011). Every day children aged between two and six play and learn in groups of 15–25 in the forest in all seasons. Usually a construction trailer (a temporary caravan-like structure used by workers on construction sites), which has been adapted to the needs of a FK (see Figure 12.1), offers shelter on cold mornings.

The children go on walks and engage in free play. Prefabricated toys are frowned upon and children are encouraged to play with the natural materials they find, engaging in imaginative play. Flora and fauna are central themes because most FKs follow a situation-oriented approach: as the children see plants, animals and the signs of seasonal change on their daily walks, these observations inform their paintings, songs or the organisation of seasonal festivals (for more details, see Schäffer and Kistemann, 2012).

Figure 12.1 Construction trailer adapted to the needs of a forest kindergarten.
Source: Schäffer, previously unpublished.

In the UK, FS developed later and in a different socio-political context from those in Germany. Similarly inspired by Scandinavian outdoor-based education, FS have been in part conceived as a kind of antidote to a perceived 'crisis' in UK childhoods (Wyness, 2000). It is argued that most British children have lost any meaningful engagement with 'nature', alongside lower levels of independent mobility and worsening educational attainment and behaviour. Thus, although in Scandinavia FS is part of conventional child education, in the UK it is a distinct practice – rarely funded directly by the UK Government (Knight, 2009).

Unlike the German context, typical FS take place over a defined timeframe – often half a day a week for six weeks – and may be offered to classes of children of all ages from a local state-run school to groups of young people (often teenage boys) with emotional, behavioural or other differences. Children engage in various activities: free play; fire-lighting; den-building; adventure/imaginative games; using knives and other tools to engage in craft activities; learning about nature (e.g. playing 'nature detectives' with a magnifying glass). Ostensibly, there were no FS until the mid-1990s, but since then there has been a sharp increase in their number – with at least 140 recorded in 2009 (O'Brien, 2009).

Care Farms (CF) are an even more recent phenomenon in the UK. There were no CF by name in the UK until 2000. The latest research shows that by 2012, there were 183 (Kraftl, 2013a). By contrast, in the Netherlands – where the best-developed network exists – there were over 1,000 CF by 2010 (Hassink *et al.*, 2010). Although there is no singular definition of a CF, in the UK they tend to fit into one of two broad types: long-standing, family-run farms that have diversified out of traditional agricultural activities and into Care Farming; or purpose-built farms – often city farms, built in city centre or suburban locations and often located near socio-economically disadvantaged neighbourhoods. In both cases, CF offer a range of activities for children of all ages, which are meant to be 'therapeutic' *and* offer learning experiences – in many cases actually privileging the latter, despite their name (Kraftl, 2014). These activities include grooming and caring for animals, horticulture, maintenance, building work and, in some cases, music- or art-based activities related to the farm's activities.

Key for all nature-based education is the (much-debated) relationship between contact with nature and dimensions of wellbeing (Kraftl, 2015). Outdoor experience forms part of the dimensions of wellbeing, as listed in the declaration of the World Health Organization (WHO) (1948), where health is defined as 'a state of complete physical, mental, and social wellbeing and not merely the absence of disease or infirmity'. More recently, the generic term 'therapeutic landscapes' (Gesler, 1992) has been deployed to describe how health-related processes emerge in different types of places. Landscapes are seen as dynamic, multidimensional constructs with physical and symbolic attributes (Gesler, 1992), some of which offer healing or care. This concept is now well embedded in health-geographical thinking (Williams, 2007) and includes hospital landscapes (Kearns and Barnett, 1999; Cooper Marcus and Sachs, 2013), but also natural landscapes (Völker and Kistemann, 2013; Gebhard and Kistemann, 2016) and their importance for health. Beyond the frame of therapeutic landscapes, many studies have examined the links between being outdoors (not necessarily in 'nature') and healthy child development (e.g. Grahn *et al.*, 1997, Kearns and Collins, 2000). For instance, some studies have shown that being outside is a stimulus for physical activity, linked with healthy feelings, lower rates of obesity (Baranowski *et al.*, 1993; Lovell, 2009) and lower rates of absenteeism due to infections (Grahn *et al.* 1997). However, from a sociological perspective, although nature may in principle offer a relatively cheap resource with salutogenic impacts (Kistemann *et al.*, 2008), children from higher socio-economic status families tend to be over-represented in nature-based settings (Powell *et al.*, 2004; Strife and Downey, 2009). Thus, one of the key themes of this chapter is an exploration of the enrolment of children in nature-based education, particularly with socio-economic class in mind.

Finally, this chapter is situated within relatively nascent social-scientific research about the FK, FS and CF. A key reason for the infancy of such research is that in contexts like the UK and Germany forest-based education and Care Farming are themselves relatively novel, albeit rapidly expanding, phenomena. Nonetheless, recent research on FS has begun to highlight children's experiences, showing that children develop greater knowledge and engagement with 'nature' (however defined) in FS than in mainstream schools (Ridgers *et al.*, 2012). It has also highlighted key skills that children acquire through unstructured play, namely teamwork, practical skills and creativity (O'Brien, 2009). Several studies have also begun to show how FS can improve happiness and psychological wellbeing (Swarbrick *et al.*, 2004; Knight, 2009). Moreover, during unstructured free play children develop independence, confidence and self-esteem (Murray and O'Brien, 2005). There are also positive connections between unstructured outdoor play and the attention spans of young children (Burdette and Whitaker, 2005; van den Berg and van den Berg, 2011), including those with attention deficit disorder (ADD) and learning difficulties (van den Berg and van den Berg, 2011).

There is very little social-scientific research about CF, especially from contexts outside the Netherlands (Hassink *et al.*, 2010). Some research has been devoted to the historical contexts of Care Farming. A key driver has been the *socialisation* of care – its relocation from institutions (like asylums or care homes) to community or family-based institutions, as part of the rolling-back of state services (Hassink *et al.*, 2010). A second key context has been a series of challenges to more traditional farming activities, especially in the UK – including the BSE (Bovine Spongiform Encephalopathy) and foot and mouth crises, and financial pressures enforced by demands for cheap food – which has encouraged farmers to look for new income streams. As with FS/FK, emergent evidence has shown that working with animals can have positive outcomes for (young) people with emotional or behavioural differences, enabling them to relate to other people in ways they were not able to previously (Berget and Braastad, 2008). Moreover, Hine *et al.* (2008) have argued that the kinds of learning outcomes found at CF are similar to those found at other 'alternative' education spaces – again, noting outcomes like increased confidence, self-awareness and empathy (Hine *et al.*, 2008). Nevertheless, unlike this chapter, very few studies examine the perspectives of either practitioners or clients at CF (except Kraftl 2013a, 2014).

Links between 'natural' education and healthy education in Germany and the UK

This section outlines the two separate studies reported in this chapter. The first examined FK in Germany. The second focused (in part) on FS and CF in the UK.

Forest Kindergartens in Germany

In 2009, 13 FK in Germany were visited and data collected through informal observation sessions of 5–9 hours in each centre, alongside qualitative interviews with teachers. The interviews were recorded, transcribed and then analysed with Atlas.ti. Data collection and subsequent analyses focused on the following themes: structure and organisation of a German FK; the everyday life of FK teachers; and the teachers' views on the synergies between long-term nature experience and health. The study analysed one FK in detail to identify its essential characteristics, and then correlated this with the other 12 FK to give an insight into the character of all German FK.

Forest Schools and Care Farms in the UK

The five FS and ten CF visited in the course of the UK research were included within a much longer-standing and more wide-ranging programme of research from 2003 onwards. The research (published in Kraftl, 2013a) involved visits by Kraftl to 59 alternative education spaces in the UK, which also included Steiner, Montessori, Democratic, Human-Scale and Home-School spaces. At each site, the author visited for a period of between one and three days, undertaking observant participation, engaging children, adults and parents in informal conversations, and carrying out 114 formal semi-structured interviews with learners, parents and practitioners. These data were (when permission was granted) audio-recorded, transcribed and combined into a single dataset whereupon they were subject to thematic analysis.

A comparative analysis of nature-based education spaces

The remainder of this chapter consists of a comparative analysis of findings of these studies. There are many overlaps between the two research projects; however, it should also be noted that because the two studies were undertaken independently, they inevitably had different aims and foci. In this chapter, the German study reports more directly on nature experiences and health; the UK study focused on geographies of alternative education, meaning that reference can only be made to *possible* health outcomes in discussions about care, emotion and embodied skills. Nevertheless, we identified three key themes where there are important points of comparison between the two contexts. We begin by focusing on how children come to attend a FK, FS or CF, before focusing on social background, health, wellbeing and education.

Starting at a Forest Kindergarten, Forest School or Care Farm

In both the German and UK contexts, children may begin attending nature-based education on the recommendation of a professional practitioner. In Germany, health is a more overt selection criterion: paediatricians may recommend to parents that children attend a FK if they have allergies, perception disorders or attention deficit disorders. However, German children cannot be *referred* to a FK as they might be to a hospital. Ultimately, it is a private decision made by parents. Furthermore, state-funded day care is rare in Germany, especially for children under three years old. Day care for children aged three to six is under municipal, church or private sponsorship. FK are mostly privately funded and waiting lists are common.

Parents' reasons for sending their children to a FK vary; health reasons (like lower infection rates) are present but not necessarily foremost:

> Sophie, FK teacher: To parents, to us adults, being outside often means rest and recovery. To gather strength, to relax, to stop thinking so much, to free

our minds, even to get some exercise through walking or cycling. To remove ourselves from our working lives and from pressure and to relax. . . For parents or for any adult that means, 'I'm going outside'. You can call it health, alleviation, mental alleviation, or whatever, I don't really know the right term for it. But I believe that we want to enable children to have this, on the basis that I can occasionally get outside, and my child can be outside every day.

Stella, FK teacher: Yes. I believe this [health] is one selection criterion. So not crucial, no, I don't believe it is. I think it's more nature that is crucial here and the staff–student ratio. As I say sometimes, we have our own little island here.

In the UK, however, children *can* be referred by a professional practitioner to either a CF or FS. Significantly, however, health is not usually the predominant concern; rather, children may be referred by a schoolteacher, educational psychologist or other local state representative based on their *educational* needs. For instance, Ursula explained the variety of ways children could be referred to her CF alone:

School referrals if they are having difficulties with the academic side, so they do an alternative vocational curriculum. [Also] we've always worked with special educational needs schools in the area. Referrals from care workers that have used us as a facility before. And we have visits from local primary schools, for day-trips. The list goes on!

Compared to Germany, health is a less prevalent reason for referral to a FS or CF in the UK. Despite their name, even CF are viewed predominantly as *education* spaces (Kraftl, 2014) rather than therapeutic spaces/landscapes. There are therefore both similarities and differences between the two countries in the process of enrolling children at nature-based education spaces. The key difference, however, appears to be the significance afforded to 'health', which is downplayed in the UK compared to Germany.

Social background and questions around equality of access

Equality of access to nature-based education spaces is an important consideration, especially in terms of socio-economic class. In Germany, several teachers reported that parents who were professionals were over-represented amongst parents in a FK. As Sophie attested:

Parents made a completely informed decision to bring their child here and that also means that they are interested in their children's education. Um, that tells you that they themselves are relatively well educated parents.

Several other teachers characterised families at their Kindergarten as (upper) middle-class. However, three of the 13 teachers reported mixed social backgrounds.

Despite the influence of parental occupation, pragmatic considerations seem to be key to parents' decisions: most FK close at lunchtime and therefore parents must have flexible working hours (e.g. self-employed), be full-time parents or have additional childcare arrangements. Since 2009, however, the German government has placed greater emphasis on mothers entering paid employment, and therefore many German day-care centres (including FK) have extended their operating hours and are also open for children under three years old. This move may reduce class-based inequities of access.

In the UK, the picture is rather different. This difference is not the result of any UK-specific impulse to make nature-based education more 'inclusive'; rather, it is indicative of how the classed politicisation of the family (and of childhood) has extended to nature-based education spaces. Lower-income families and their children have often been identified as a cause of the 'crisis of childhood' (Wyness, 2000; Cobb, 2007). This un-evidenced assumption has led to a range of interventions into the lives of disadvantaged children, focusing most recently on so-called therapeutic interventions into children's emotional wellbeing (Ecclestone and Hayes, 2008; Conroy, 2010). The vast majority of FK and CF in the UK study therefore catered for children from disadvantaged backgrounds. In some cases – as for a CF on the edge of a deprived suburb in England – this was their reason for existing:

> Jenna, CF practitioner: It's for the local community, initially. Aiming at two lower income housing estates. But then it's spread further than that. Now we can have anyone come through our gates as we have an open-gate policy. So we can have kids from the local [deprived housing] estates come as they please, a local school group, a group of kids with special educational needs, all at the same time.

Jenna's experience highlights the diversity of FS and CF in the UK; however, it also evidences the initial objective of many such spaces: to cater for disadvantaged communities. Many FS and CF are therefore located near such communities, or specialise in providing education for children at risk of exclusion from school (who in turn predominantly come from lower-income families). Indeed, it is such specialisation that enables FS and CF to continue to attract funding through grants and charitable donations.

How do interventions around 'health' and 'education' overlap in nature-based education spaces?

As discussed in the introduction to this chapter, the WHO's definition of health includes physical, mental and social aspects of wellbeing. In this section, we focus on each in turn, before acknowledging the entanglement of these three facets of wellbeing in teachers' and parents' talk about health and education.

Physical wellbeing

In the German study, most discussions of health concentrated on physical activity, coordination and the impact of outdoor play on the immune system. For example, two FK teachers Olivia and Kathrin noted:

> Olivia: Yes, motor skills, simple muscle strengthening, coordination – hand to foot, hand to eye, whatever kind of coordination. Simply being able to move, even on difficult terrain.

> Kathrin: And the opportunities for movement are simply different here. . . than in those crowded rooms, where even the gym is used by 50 children at once. That is just crazy.

In the UK, forests and other 'natural' spaces are similarly viewed as places where children have more space to move. At one FS, where a local primary school took their children for occasional morning visits, Debbie said:

> The boys – and occasionally the girls – are just so confined in the classroom. The forest gives them room to breathe, to expand, to shout, without being on a tight rein. It must be good for them, even if it's just the change of scene.

Indeed, many FS practitioners – who were passionate about woodland environments – also conceded that it was sometimes the simple act of moving a young person from their routine everyday environment to another that could accrue a range of benefits:

> Joanne, FS practitioner: Actually, it's not just FS – the thing about the sessions is that you *are* remote from the community, physically and, you know, mentally. That you *have* walked a bit, that you have an opportunity to explore, becoming confident with your body in a totally new space.

Mental wellbeing

Facets of physical health mentioned previously overlap with mental health. In both the UK and German contexts, respondents' reflections on mental wellbeing echoed powerful discourses about the benefits of outdoor education (discussed in the introduction). For instance, echoing the tenets of alternative education more broadly (Kraftl, 2013a), positive impacts for mental wellbeing include games played according to children's interests at their 'own pace' without sensory overload. In this regard, the German and UK contexts were remarkably similar:

> Sophie, FK teacher, Germany: But also we have the entire morning, and can decide ourselves on the pace, how quickly we want to do things . . . and also direct how intensely we want to do things. And, um, I think it is simple, when

children are learning that they can make their own decisions and get settled on a task. They know that 'they can slow down a process, they can continue at this intensity level, and investigate, and look, and turn around and measure something again and so on'.

Charlotte, home-schooling mother whose children attend a FS in the UK: It's just not having it contrived – oh we'll go to the park and play now. It's having that space to sit and get bored, and then start reflecting on things. Just sitting for hours in a tree and staring. I think that's important emotionally.

Although neither study explicitly asked about silence, both Sophie and Charlotte – like several other interviewees in both countries – implied that lack of noise led to a range of psychological and/or emotional benefits (in this regard, see Conradson, 2007). In Germany, several teachers hinted at how FK were characterised by low noise levels and limited sensory stimuli, in contrast to regular kindergartens:

Maria, FK teacher: It just does them good. They are so relaxed.

Similarly, in the UK the vast majority of teachers and parents talked about how a key characteristic of FS and CF was that they could make children more relaxed, serene and considered (Kraftl, 2013a).

In the German study, there was greater discussion of the direct role of 'nature' in producing such embodied changes within children; in the UK, less so. The educational focus in FK is the forest as a living space and the careful handling of nature. Everyday perception of, and occupation with, nature is central to children's learning *and* health-giving experiences:

Ruth, FK teacher: So it is the everyday handling of what they find . . . That's just it. We also deal with a specific curriculum. But I think it is natural for them to see what they find there and use it as a starting point for what they can get out of it. It is better for children than if I am always trying to engage them in a topic, to get my agenda into their heads.

Perhaps the most significant finding, however, is that in Germany, as in the UK, the emphasis is not simply on placing children 'in nature' and, in a deterministic way, waiting for nature to have an effect. Rather, a FK education proceeds through a series of carefully planned activities wherein the 'natural' and the 'social' are *combined* in heterogeneous, dynamic assemblages of what have been termed 'hybrid childhoods' (Kraftl, 2013b) or 'place-responsive pedagogies' (Mannion *et al.*, 2013). This could be through research projects into food chains (frog spawn, fish, heron) or creative crafts with natural materials. Sophie (FK teacher) explains one creative roleplay:

We hang like bats. The children hang in a climbing frame on the playground like bats with their heads down, yes. And we see how it is when they spread their wings, like a skeleton, like fingers, with wings stretched across them. So they are similar to humans. Yes, such things happen, if everyday life suggests it . . .

Thus on the one hand, mental wellbeing is supported by natural spaces that children can modify to meet their own needs in their own time. On the other, outdoor educators do not see a simple, deterministic relationship between nature and children's health; rather, 'nature' only works through carefully planned activities that may improve mental wellbeing.

Social wellbeing and holistic health

The holistic WHO definition of health emphasises physical, mental and social wellbeing. Significantly, this definition bears striking similarities to the pedagogical aims of nature-based education. In Germany, in contrast to mainstream preschool environments, teachers aim at forms of *cooperative* social interaction between teachers, pupils and parents.

> Doris, FK teacher: The children have a really strong group spirit, because we walk through thick and thin with each other in every type of weather.

In both the UK and Germany, practitioners consistently underlined the importance of rituals and rhythms in structuring children's experiences:

> Florence, FK teacher, Germany: And the morning circles are very important. . . . Because they create a group spirit. These are regular events in the daily routine.

> Paula, FS teacher, UK: We always start off with the same activity, each week. And we set boundaries, which may get bigger over time, allowing more freedom, but it's important to have set boundaries, through the day and in the wood, so that we can try to get a common bond going in the group. And it doesn't always work – we only have them for a period of six weeks, one day a week [. . .] in fact it rarely works as we'd like to see it.

In Germany, several teachers highlighted how language is a crucial element of social interaction in FK. As Ruth noted:

> They can all explain things really well. . . . And describe things also, because naturally in the forest, they have to explain things they play with in another way. . . . So that their partner knows what for instance this stick is now. Is it a car or a plane or a magic wand or . . .?

However, in the UK, although they did not downplay the importance of language, several CF practitioners noted the importance of developing shared bodily skills and habits (such as gestures and facial expressions) through which young people could demonstrate empathy, compassion or a common group identity (Kraftl, 2013a). Aspects of social wellbeing, such as care for others, compassion, empathy and collaboration, were emphasised in nature-based education spaces in both countries.

The level of parent participation in the nature-based educational settings varies between countries. In Germany there is a high level of involvement of parents (with many FK organised as parental initiatives), whereas in the UK parental involvement in FS and CF is, by comparison, lower.

In this section we have identified differences in how nature-based education is organised and experienced in UK and Germany with particular reference to aspects of health. The distinctions emerged as much from the flow of questions/answers in the interviews in our research as from distinctions drawn by respondents. More importantly, when they discussed health, most respondents did so in a holistic way and they blurred concepts of 'health' and 'education' in complex, dynamic ways. For instance:

> Maria, FK teacher: Yes, on a physical level as well as a spiritual and mental level. . . so on all levels, I think it is just healthy for the children, all round.

> Ruth, FK teacher, Germany: Children can let out their natural urge to move. And the whole thing works positively on their bodies. So they don't have to restrict themselves or anything. Rather they can enjoy life to the full . . . And their development – I think, when the whole thing in this natural way has a positive effect on their bodies, then people are open to other things. And can use a lot more things for themselves . . . And understand more. And concentration is part of this process. So, if I got rid of my excess energy and cleared my head first, then I can concentrate better. I can absorb knowledge and the world around me . . . And I believe that children who can act out their fantasies are much healthier.

From her observations of children in FK, Ruth postulates that a capacity for concentration is enhanced by intensity of movement and together they contribute to the balanced character of the children.

Conclusion

This chapter has focused on a selection of the aspirations, objectives and planned outcomes of FK, FS and CF. It has emphasised some of the benefits of a cross-cultural comparison of nature-based education spaces. Three key findings emerged from our analysis. First, although there are similarities in the principles behind FK, FS and CF in Germany and the UK, the process of enrolling new students differs markedly. In Germany, professional practitioners may recommend (to parents) that a child attend a FK, but it is ultimately the parents' decision. In the UK, direct referrals from practitioners are more common, and parents may be some of several different actors involved in decision-making processes. As we explain later, this can mean that the socio-economic status of children enrolling at nature-based education spaces differs greatly between Germany and the UK. Additionally, in Germany, the decision to enrol appears to be more based on reasons associated with wellbeing than in the UK. One key contribution of this chapter in terms of alternative education research has therefore been to highlight where and when, in the process of a young person's journey through a particular educational space, health outcomes matter *more* and, just as significantly, *less* than other considerations (compare Woods and Woods, 2009).

A second finding was that socio-economic differences are addressed differently between German and UK contexts. German FK tend to cater for a disproportionate number of children from better-off families whose parents are in professional, academic and/or flexible employment. In the UK, FS and CF tend initially to *target* young people from disadvantaged backgrounds, although they do, to a lesser extent, cater for young people from other socio-economic groups. This situation should not lead to the conclusion that the UK examples are somehow more inclusive; rather, that the (highly questionable) class politics inherent in contemporary policies and professional practices around families and young people mean the social make-up of students is quite different from Germany. This chapter has therefore begun to add further complexity to emerging debates around *which* young people are experiencing (if not benefiting from) forms of 'therapeutic', emotional and experiential education (Ecclestone and Hayes, 2008).

The third key finding is that although the focus of nature-based education spaces is education, all aspects of health and wellbeing (as per WHO's definition) are served. On the one hand, there is increasingly compelling evidence that nature plays a unique, irreplaceable role in healthy child development (Kahn and Kellert, 2002; Chawla and Cushing, 2007). The open space and silence and the contrast with children's regular home/school environments offer a holistic stimulus for health (in this regard, compare Kearns and Collins, 2000, on New Zealand-based children's health camps). On the other hand, critical studies of children in nature and of children and nature, and hybrid theorisations of childhood, seem to question pervasive contemporary wisdom about removing children from their everyday environments and placing them in 'natural' spaces (e.g. Kraftl, 2013b; Taylor, 2013). Significantly, as far as the perspectives of teachers and parents are concerned, this chapter has uncovered evidence that supports *both* arguments.

In sum, our comparative analysis has demonstrated how changing conceptions of childhood are prompting evolving conceptions of day care and education across two cultural contexts in Western Europe (Tandy, 1999; Holloway and Valentine, 2000). However, there is a clear need for future research. Most fundamentally, there is scope for international comparative studies that are planned as such from the outset using consistent, bespoke methodologies (as opposed to comparing findings from two comparable but independent studies). Equally, and in addressing the tension highlighted earlier, there is a need for interdisciplinary research that brings together theoretical and methodological approaches from psychology, neuroscience, ecology, education studies, human geography and sociology. Finally, future research might systematically address some of the tentative comparisons drawn in this chapter in terms of socio-economic class, but also in respect of other social differences, such as gender, religion, ethnicity and sexuality.

Notes

1 For simplicity, we henceforth refer to Forest Kindergarten, Forest Schools and Care Farms as 'nature-based education' when referring to these spaces in general.
2 Source: www.destatis.de/DE/PresseService/Presse/Pressemitteilungen/2014/09/PD14_313_225.html (accessed March 2014).

References

Baranowski, T., Thompson, W.O., DuRant, R.H., Baranowski, J. and Pohl, J. 1993. Observations on physical activity in physical locations: age, gender, ethnicity, and month effects. *Research Quarterly for Exercise and Sport*, 64(2), pp. 127–33.

Berget, B. and Braastad, B. 2008. Animal-assisted therapy with farm animals for persons with psychiatric disorders. *Annals Ist Super Sanita*, 47, pp. 384–90.

Burdette, H.L. and Whitaker, R.C. 2005. Resurrecting free play in young children. Looking beyond fitness and fatness to attention, affiliation and affect. *Archives of Pediatrics & Adolescent Medicine*, 159, pp. 46–50.

Chawla, L. and Cushing, D.F. 2007. Education for strategic environmental behavior. *Environmental Education Research*, 13(4), pp. 437–52.

Cobb, N. 2007. Governance through publicity: anti-social behaviour orders, young people, and the problematization of the right to anonymity. *Journal of Law and Society*, 34, pp. 342–73.

Conradson, D. 2007. Experiential economies of stillness: the place of retreat in contemporary Britain. In Williams, A. ed. *Therapeutic Landscapes*. Aldershot: Ashgate, pp. 33–48.

Conroy, J. 2010. The state, parenting, and the populist energies of anxiety. *Educational Theory*, 60, pp. 325–40.

Cooper Marcus, C. and Sachs, N.A. 2013. *Therapeutic landscapes: an evidence-based approach to designing healing gardens and restorative outdoor spaces*. Hoboken, NJ: John Wiley & Sons.

Ecclestone, K. and Hayes, D. 2008. *The dangerous rise in therapeutic education*. London: Routledge.

Gebhard, U. and Kistemann, T. 2016. *Landschaft, Identität und Gesundheit*. Heidelberg: Springer VS.

Gesler, W. 1992. Therapeutic landscapes: medical geographic research in light of the new cultural geography. *Social Sience and Medicine*, 34(7), pp. 735–46.

Grahn, P, Mårtensson, F, Lindblad, B., Nilsson, P. and Ekman, A. 1997. Ute på dagis: hur använder barn daghemsgården? Utformningen av daghemsgården och dess betydelse för lek, motorik och koncentrationsförmåga. Stad & Land, Vol. 145, Movium: Alnarp.

Hassink, J., Elings, M., Zweekhorst, M., van den Nieuwenhuizen, N. and Smit, A. 2010. Care farms in the Netherlands: attractive empowerment-oriented and strengths-based practices in the community. *Health and Place*, 16, pp. 423–30.

Hine, R., Peacock, J. and Pretty, J. 2008. Care farming in the UK: contexts, benefits and links with therapeutic communities. *Therapeutic Communities*, 29, pp. 245–60.

Holloway, S.L. and Valentine, G. 2000. *Children's Geographies: Playing, Living, Learning*. London: Routledge.

Kahn, P.H. and Kellert, S.R. 2002. *Children and nature: psychological, sociocultural and evolutionary investigations.* Cambridge, MA: MIT Press.

Kearns, R.A. and Barnett, J.R. 1999. To boldly go? Place, metaphor and the marketing of Auckland's Starship Hospital. *Environment and Planning D: Society and Space*, 17(2), pp. 201–26.

Kearns, R.A. and Collins, D.C.A. 2000. New Zealand children's health camps: therapeutic landscapes meet the contract state. *Social Science and Medicine*, 51, pp. 1047–59.

Kistemann, T., Claßen, T. and Schäffer, S. 2008. Naturschutz und Gesundheitsschutz. Identifikation gemeinsamer Handlungsfelder. In: Erdmann, K.-H., Eilers, S., Job-Hoben, B., Wiersbinski, N. and Deickert, S. eds. *Naturschutz und Gesundheitsschutz: Eine Partnerschaft für mehr Lebensqualität*, pp. 25–34.

Knight, S. 2009. *Forest schools and outdoor learning in the early years.* London: SAGE Publications.

Kraftl, P. 2013a. *Geographies of alternative education: diverse learning spaces for children and young people.* Bristol: Policy Press.

Kraftl, P. 2013b. Beyond 'voice', beyond 'agency', beyond 'politics'? Hybrid childhoods and some critical reflections on children's emotional geographies. *Emotion, Space and Society*, 9, pp. 13–23.

Kraftl, P. 2014. Alternative education spaces and local community connections: a case study of Care Farming in the UK. In: Mills, S. and Kraftl, P. eds. *Informal education, childhood and youth: geographies, histories, practices.* Basingstoke: Palgrave, pp. 48–64.

Kraftl, P. 2015. Alter-childhoods: biopolitics and childhoods in alternative education spaces. *Annals of the Association of American Geographers*, 105, pp. 219–37.

Lovell, R. 2009. *Evaluation of physical activity at Forest School.* PhD thesis, University of Edinburgh, UK. Available at https://www.era.lib.ed.ac.uk/handle/1842/4146 (accessed 19 February 2017).

Mannion, G., Fenwick, A. and Lynch, J. 2013. Place-responsive pedagogy: learning from teachers' experiences of excursions in nature. *Environmental Education Research*, 19, pp. 792–809.

Miklitz, I. 2011. *Der Waldkindergarten. Dimensionen eines pädagogischen Ansatzes.* Berlin: Cornelsen Scriptor.

Murray, R. and O'Brien, L. 2005. *'Such enthusiasm – a joy to see'. An evaluation of Forest School in England.* Forestry Commission. Available at www.forestry.gov.uk/pdf/ForestSchoolEnglandReport.pdf/$FILE/ForestSchoolEnglandReport.pdf (accessed 19 February 2017).

O'Brien, E. 2009. Learning outdoors: the forest school approach. *Education*, 3(13), pp. 37, 45–60.

Powell, L.M., Slater, S. and Chaloupka, F.J. 2004. The relationship between community physical activity settings and race, ethnicity and socioeconomic status. *Evidence-Based Preventive Medicine*, 1(2), pp. 135–44.

Ridgers, N., Knowles, Z. and Sayers, J. 2012. Encouraging play in the natural environment: a child-focussed case study of forest school. *Children's Geographies*, 10, pp. 49–65.

Schäffer, S. and Kistemann, T. 2012. German Forest Kindergartens: healthy childcare under the leafy canopy. *Children, Youth and Environments*, 22(1), pp. 270–9.

Strife, S. and Downey, L. 2009. Childhood development and access to nature. *Organization & Environment*, 22(1), pp. 99–122.

Swarbrick, N., Eastwood, G. and Tutton, K. 2004. Self-esteem and successful interaction as part of the forest school project. *Support for Learning*, 19, pp. 142–46.

Tandy, C.A. 1999. Children's diminishing play space: a study of inter-generational change in children's use of their neighbourhoods. *Geographical Research*, 37(2), pp. 154–64.

Taylor, A. 2013. *Reconfiguring the natures of childhood*. London: Routledge.

van den Berg, A.E. and van den Berg, C.G. 2011. A comparison of children with ADHD in a natural and built setting. *Child: Care, Health and Development*, 37(3), pp. 430–9.

Völker, S. and Kistemann, T. 2013. 'I'm always entirely happy when I'm here!' Urban blue enhancing human health and well-being in Cologne and Düsseldorf, Germany. *Social Science & Medicine*, 78, pp. 113–24.

Williams, A., ed. 2007. *Therapeutic landscapes*. Aldershot: Ashgate.

Woods, P. and Woods, G. 2009. *Alternative education for the 21st century*. London: Palgrave.

World Health Organization (WHO). 1948. Preamble to the Constitution of the World Health Organization as adopted by the International Health Conference, New York, 19 June–22 July 1946; signed on 22 July 1946 by the representatives of 61 States (Official Records of the World Health Organization, no. 2, p. 100) and entered into force on 7 April 1948. The definition has not been amended since 1948.

Wyness, M. 2000. *Contesting childhood*. London: Routledge.

13 Being connected?

Wellbeing affordances for suburban and central city children

Christina R. Ergler, Robin Kearns,
Adrienne de Melo and Tara Coleman

The ways in which urban children understand and interact within their local environments has changed dramatically over recent decades. Where once children roamed free with a permissive licence to explore, increasingly there are constraints on their mobility and restricted exposure to less contrived and controlled spaces (Gleeson and Sipe, 2006). This change has origins in both the organisation of, and conduct within, cities. Auto-dominance, urban compaction and the trend towards vertical living, as well as the dominance of indoor settings (e.g. the private home, shopping malls and gaming arcades), are often identified as encouraging this development (Freeman and Tranter, 2011). A related influence is the increase in safety-conscious parenting practices such as 'chauffeuring' that have arisen from concerns with heightened traffic volumes and 'stranger danger' (Gleeson and Sipe, 2006).

As a result of the foregoing processes, researchers, politicians and play advocates have voiced concern that children residing in Western cities are increasingly losing the opportunity to develop familiarity with local and especially 'natural' environments (Louv, 2005; Malone, 2007). In this chapter, we explore central city and suburban children's knowledge and familiarity with their local environment under the umbrella of 'wellbeing affordances'. For us, as we detail later, wellbeing affordances consist of children's sense of place (knowing a place), place attachment (a feeling of belonging to a place) and children's spatial literacy of social and physical environments (Relph, 1976; Chawla, 1992; Freeman and Vass, 2010). We postulate that a positive relationship with, and feeling well-connected to, local environments is important for children's wellbeing on an individual level. Furthermore, this relationship can potentially contribute to fostering and maintaining healthy urban (green) places that, in turn, contribute to planetary wellbeing. In making these links, we bring into convergence two ideas: that the health of ecosystems largely depends on local residents' understanding of their character; and that models of health increasingly embrace the natural environment as a key to wellbeing (Spencer and Blades, 2006; Parkes *et al.*, 2010; Parkes *et al.*, 2011; Davis and Elliott, 2014). The state of the environment and the exposure to natural settings determines people's health and wellbeing.

Various writers contend that, for children, spending time in more pristine natural settings is associated with diverse physical health benefits and positive attitudes towards the environment (Kahn, 2002; Wells and Lekies, 2006; Gill,

2014). However, even being in manicured green spaces can positively influence children's health through encouraging increased physical activity levels (Timperio *et al.*, 2004), while children participating in gardening projects and engaging with elements of the natural food supply tend to have healthier eating habits (Morris and Zidenberg-Cherr, 2002). Other studies highlight the restorative aspects of spending time within proximity to natural environments for children's mental and emotional wellbeing (Faber Taylor *et al.*, 2002; Wells and Evans, 2003). Being engaged with, and aware of, the components of local ecosystems has been shown to encourage a commitment to local environments and eco-friendly attitudes that are carried into adulthood (Kahn, 2002; Wells and Lekies, 2006). Although the health and wellbeing implications of children engaging with nature in their suburban surroundings have been considered (Freeman *et al.*, 2015), less is known about central city children's awareness of their local environment and how both wild and manicured green spaces and iconic local places can contribute to their wellbeing.

In this chapter we explore how the connection to 'natural' and mundane local spaces differs between suburban and central city children, and what implications this has on children's wellbeing affordances. For us, wellbeing affordances encompass all physical, social and mental activities available in local environments that are potentially therapeutic by fostering feelings of wellbeing and/or attachment to place. In other words, we consider how ordinary local environments offer wellbeing affordances for primary school age children by linking the concepts of environmental literacy, affordances and a sense of place (Relph, 1976; Chawla, 1992; Freeman and Vass, 2010). We explore suburban and central city children's experiences of these components in contributing to their individual wellbeing and the health of the environment.

In the following sections we outline our conceptualisation of wellbeing affordances in order to contextualise our analysis of urban children's wellbeing. We then present case study evidence of how suburban and central city children connect to the social and physical fabric of their neighbourhoods and what features and activities they value as part of feeling and being well in their neighbourhood. We conclude that a recursive relationship between socio-spatial literacy, a feeling of wellbeing and being attached to neighbourhood places matters both on the individual and global scale.

Situating wellbeing affordances, environmental literacy, affordances and a sense of place

The starting point for our conceptualisation of 'wellbeing affordances' is the idea of 'environmental literacy'. On the one hand, the term encompasses children's spatial awareness and knowledge about their social and physical surroundings (Hart, 1979; Blaut, 1987; Freeman and Vass, 2010). Children 'read' their environment in terms of its key environmental, social and cultural elements (Malone, 2007). Such an understanding also recognises that children's capabilities and confidence in navigating environments grows through forming relationships with, and attachments to, their surroundings. On the other hand, the term 'environmental literacy' can also be read beyond the notion of socio-spatial literacy. In environmental education the term 'environmental literacy' became popular in the 1970s and emerged from concerns about the management of environmental resources (Roth, 1992). Researchers saw a need to educate people on how human interactions with the environment were (and are) affecting planetary wellbeing (Berkowitz *et al.*, 1997; McNeill and Vaughn, 2012). In this chapter we are concerned with both aspects of environmental literacy: the development of ecological awareness and, through emerging sense of place, sympathetic protection activity. We speculate that such activity can contribute to wellbeing in at least two ways: it can be physically and mentally healthy at the level of the person, and it can also be protective of the (broadly-sketched) health of the local environment.

To examine children's environmental awareness and literacy many earlier studies have focused on children's reconnection with 'wild' nature (e.g. Louv, 2005; Gill, 2011). In this study we take children's informal play from wild places (e.g. national and regional parks) to the neighbourhood, and explore how local wild places (aka the 'bush' in the New Zealand context) and manicured 'natural' environments (e.g. flowerbeds, trees on side of road), as well as mundane everyday spaces such as wide sidewalks or local shops, may offer children 'wellbeing affordances'. To address the diversity of play and engagement opportunities that mundane everyday spaces offer children for fostering a feeling of being well, we turn to the concept of 'affordances' from ecological psychology (Gibson, 1979; Kytta, 2002).

'Affordances' call attention to the functional properties of places or objects (e.g. a park bench can afford sitting, climbing or jumping), which structure or foster children's interactions with a local environment. Objects such as a tree or statue stimulate and generate awareness of the possible actualisation of their environmental properties (Heft, 1988). Environments – be they 'natural' or constructed – are therefore more than the simple backdrops in which children's activities occur or are absent from. Instead, children transform environments and objects for *their* play needs, confirming that such environments can be regarded as 'enabling spaces' (Duff, 2011) in the sense that the place itself tacitly offers some agency in offering opportunities for creative activity. This view implies on the one hand that any place can be an adventure playground, but on the other hand suggests that play practices are place-and context specific (Lim and Calabrese Barton, 2010).

What the child perceives as affordances in an environment are 'constructed and composed' out of 'diverse social, affective and material resources' (Duff, 2011: 155). The 'where' and 'when' of play affects the 'how' of this core 'work' of childhood. Social norms and rules (e.g. parenting cultures, see Dowling, 2000), as well as the availability and accessibility of play locations, and how children perceive, evaluate, transform and make sense of environments for their play needs all influence where, when and how play affordances are normalised and actualised. In turn this means that, for us, wellbeing affordances consist of a complex triad of social norms and societal expectations, children's environmental literacy and their affection for affordances in different places. The link between knowledge of place and the emotional ties children develop through repeated experience of them is central for our conceptualisation of wellbeing affordances of urban children.

To further embed the links between emotions, affect and place, theoretically we draw on the geographical concept of 'sense of place' (Eyles and Williams, 2008; Bartos, 2013). This concept involves a link between knowledge of place and the emotional ties people develop through repeated experience of positive feelings in places (place attachment). The term 'place', therefore, does not simply refer to a geographic location (Cresswell, 2014); rather, it suggests that the meaning of place, which evokes feelings of awe, attachment and belonging, enhances the values attributed to local environments – turning a space into a place. To conceptualise this complex relationship, Tuan (1974) established the idea of 'topophilia', which describes the affective bond between people and a place and encompasses the generation of place-focused feelings of appreciation and care. Being within a place therefore leads to belonging to, identifying with and feeling rooted in place (Relph, 1976; Tuan, 1977; Hernandez *et al.*, 2007).

It is not only the outcome of feeling attached but the process of becoming attached (Harris *et al.*, 1996) that is important because this process draws attention to the individual's explanations for possessing particular place-related emotions, attachments and subsequent behaviours (e.g. caring for or protecting a place). As children actualise, experience and transform local affordances and generate positive socio-spatial interactions, their sense of place is therefore heightened over time. They produce affective socio-spatial bonds growing up in and utilising their local environment. In this sense, the concept of wellbeing affordances suggests that attention should be paid to how people feel towards particular places (e.g. the affective dimensions of attachment and experience; see also Hidalgo and Hernandez (2001), and Cresswell (2014) for an adult perspective); and whether feelings towards, and involvement in, place are generative of values directed at protecting a place (e.g. the development of pro-environmental attitudes that may be carried over into adulthood or environmental activism; Wells and Lekies, 2006).

In the following sections we show how suburban and central city children negotiate wellbeing affordances within the limits of wider societal norms and rules. We consider children's experiences of commonly known child spaces (e.g. parks and playgrounds) and what might otherwise be overlooked spaces of urban life, such as flowerbeds or sidewalks. We argue that children's positive experiences of these elements of the urban environment and engagements with other human beings afford opportunities for fostering physical, social and mental wellbeing on the individual level and the health of the environment across all scales.

We draw on go-along interviews (Loebach and Gilliland, 2010) with ten participants, which were recorded verbally with a digital recorder and visually through global positioning systems (GPS) and photos taken by the children. In addition, adults accompanying the children on the walks observed and noted the activities. Local children showed non-local children around their neighbourhood and the latter engaged in and questioned them about locally appropriate play activities. Children were instructed to show the non-local children fun places to play and activities or localities that make them happy. Talk generated on the walks was transcribed and, together with the photos, coded and analysed thematically. The participating ten-year-olds lived in either a suburban neighbourhood (detached

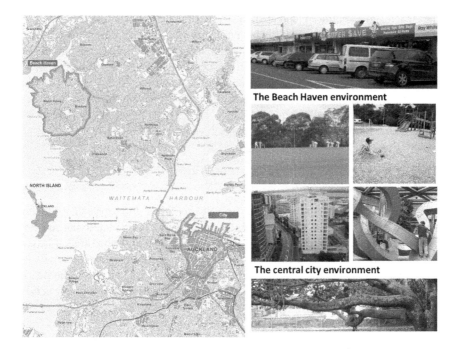

The Beach Haven environment

The central city environment

Figure 13.1 Overview study locations and environmental differences.

houses with gardens) or central city location (high-rise apartment complexes without outdoor communal spaces or playgrounds) in Auckland, New Zealand (population in 2013 of 1.4 million; Statistics New Zealand, 2013). For more details on the method, see Ergler (2011).

The suburb of Beach Haven and Auckland's central city are both lower socio-economic neighbourhoods with distinct physical boundaries. Beach Haven has numerous parks, some of which include courts or sport fields (see Figure 13.1). The suburb is well-known as a 'good' place to raise children because community ties are strong and well developed (McCreanor *et al.*, 2006). The central city is Auckland's major economic hub and residential high-rise living is a relatively recent phenomenon. Apartments were designed to cater for students or empty-nesters (Carroll *et al.*, 2011) and have limited communal outdoor space. Resident city children can play in two major parks (Myers, which has the only playground, and Albert Park) and in some smaller landscaped green spaces within the central city boundaries (see Figure 13.1).

Viewing local wellbeing affordances through the eyes of suburban and central city children

Based on the places the children identified as aspects of their neighbourhoods that were most important to them and that they showed each other on the walks, the photographs they took and the comments they made along the way, it appears that the Beach Haven children tend to spend more time in more varied settings within their neighbourhood (e.g. playground, park, netball courts, gardens, street, wharf) that foster physical, social and mental wellbeing. Places designed with children in mind are limited to a few locations in the central city (e.g. playground, library, after-school care facilities). Nonetheless, all participating children demonstrated diverse socio-spatial knowledge of their respective local environments and they revealed numerous wellbeing affordances of their neighbourhoods, which we discuss next for the suburban and central city locations.

Children's perceptions and experiences of Beach Haven's wellbeing affordances

The suburban children pointed out many of the social and physical wellbeing affordances of Beach Haven. In all their accounts there was an underlying sense of pride for their neighbourhood. For example, all children described Beach Haven Wharf (an old ferry landing used for swimming), the local beach and park with its bush area, sports fields and playground as unique and 'special places'. The Beach Haven children tended to value the natural environment most within their neighbourhood because it forms part of their sense of place (e.g. Hernandez *et al.*, 2007). Children's photos in the suburb were dominated by natural features (e.g. trees, flowers; see also Figure 13.2). Maree[1] (BH), for example, explained why they took so many photos of natural settings. She said that Beach Haven is a good place for children to live: 'our landscape out here is really good. There is a lot of greenery'. This is an observation consistent with our conceptualisation as 'green' wellbeing affordances. Another girl repeatedly took photographs of flowers and birds because she believed they symbolise her neighbourhood and are features that can be observed as a leisure activity: 'I really need a picture of a bird . . . and I've got lots of photos of flowers, because they're nature and our neighbourhood has a lot of nature (compared to others)' (Nicole, BH). Children frequently pointed out that other residential areas lack so many open and easily accessible green and blue spaces, effectively individuating their suburb from others in terms of 'topophilia' (Tuan, 1974) and setting the area apart in terms of available wellbeing affordances.

Figure 13.2 Children's interaction with the natural environment during the neighbourhood walk in Beach Haven (climbing a tree, picking honeysuckle flowers and drawing on a beach). (Children took all photos in this section.)

In the eyes of the suburban children, natural spaces afforded physical and mental health and wellbeing. For example, Michael (BH) mentioned that the natural environment is a place in which to both play and relax. He showed a central city child the bushy area where he frequently collected insects, leaves and stones and a lookout point across the estuary. Michael seemed to find retreat in this natural environment and it provided a place where he could 'recharge his batteries' before he wandered off to the next play adventure. This anecdote suggests that natural environments have an important role in identity-formation while offering a place in which to play and facilitate therapeutic values (Moore, 1986; Williams, 2007). In terms of this chapter's key theme, Michael (BH) finds a sense of place (literally a haven) within Beach Haven's bush and beach settings. This feeling facilitates an environmental literacy from which he can 'read' the landscape for its opportunities for play (activity) and rest (therapy), which we conceptualise in their combined form as wellbeing affordances.

Exploration and play within natural environments can foster feelings of wonder and a sense of ownership (Wilson, 1997; Mergen, 2003), observations that are apparent from Beach Haven examples. Some of the Beach Haven children's interactions within the natural environment (e.g. climbing trees, picking flowers and carving names into rocks at the tidal beach, see photographs in Figure 13.2), suggested that they were relatively well connected to the nature within their neighbourhoods. Children are known to seek natural places for play (Moore, 1986) and this opportunity at Beach Haven appeared to contribute positively to their knowledge about their neighbourhood and environmental protection. Children, including Nicole (BH), expressed personal concern for the natural environment, saying 'Some people don't care about the environment . . . I find that really sad, they just dump things', and others pointed out the beauty of native plants and their worthiness and need for protection (see Figure 13.2 trees in background: 'That's a Nikau palm over there . . .' (Maree, BH)). We suggest, therefore, that exploration and play can foster a particular sense of place and identity. In other words, knowledge of the natural environment may lead to pro-environmental attitudes and affinities and, ultimately, a stronger sense of place and identity through a feeling of de facto ownership and personal responsibility (see also Wells and Lekies, 2006). As these examples illustrate, the connection to the local environment affords an engagement with the health and wellbeing of the local ecosystem.

The countervailing view that central city children may be less accustomed and familiar with the natural environment also became evident through the Beach Haven children's actions. For example, Maree and Isabelle (BH) collected honeysuckle flowers for the central city children, showing awareness that there are plants growing on the street from which nectar can be tasted (see Figure 13.2). These children clearly had an environmental literacy that extended to 'reading' the landscape for its distinctive sensory qualities, and were sufficiently excited by knowing its taste as well as more obvious sights. They sought to ensure that this knowledge, the wellbeing affordances of Beach Haven, not only embraced the visual beauty of the flowers, but also their aesthetics for taste buds.

Arguably, such environmental knowing comes from a competency won by a set of permissive parental licences, suggested by comments such as: 'I walk to school every day . . . it's close to home' (Michael, BH) and 'I'm allowed to go to the shops alone, or I go with my mates' (Valerie, BH). Similarly, the majority of suburban children like Josh (BH) said that they go to Shepherd's Park regularly after school (see Figure 13.1: sport fields). This is one of the largest parks and a meeting hot spot for unsupervised children; it is a popular recreation venue,[2] located approximately 1 km away from Beach Haven Primary School. Josh and Michael (BH), for example, mentioned that they often went on bush walks in Shepherd's Park and other local areas. They both explained that they felt safe in these natural environments because there were always other people around. The boys therefore felt confident engaging with natural settings and could rely on people in the community in Beach Haven if something went wrong. Their story highlights the recursive relationship between environmental literacy, sense of place and wellbeing. The children can 'read' the landscape, knowing its cues for safety and having the security of others within reach and, through engagement in these 'off-the-beaten-track' spaces, both extend and confirm their experience of wellbeing.

During the walk, it became clear that one of the children from Auckland Central seemed quite uneasy amidst Beach Haven's natural environment. He was unable at this point to read the wellbeing affordances BH children saw and valued about their neighbourhood. For McBeth (CC), 'This sand feels uncomfortable . . . It's too muddy' and 'I hate the slippery sand! I wish it was just like plain concrete stairs so I don't fall down'. He later exclaimed 'Watch out, don't fall into the water . . . you could break your head on the rocks . . . and there could be sharks in there!' McBeth's (CC) description of concrete stairs as 'plain' signals his familiarity with, and preference for, the built environment. This was also shown through his requests to go into the shops and buy things during the neighbourhood walk. For him, going into the shops appeared to be a form of play (see also Ergler and Wood, 2015). This preference for consumption over creative engagement with the natural environment suggests that McBeth (CC) had a greater sense of place and self within adult commercial spaces, as opposed to natural settings, a view reinforced by his central city neighbourhood walk.

Children's perceptions and experiences of Auckland Central's wellbeing affordances

From the neighbourhood walk in Auckland Central, it was evident that child-friendly spaces are scarce and that the performance of childhood is mainly an indoor experience shaped largely by auto-dominance and parental fears of neighbourhood safety. For example, during the walk, Taesong (CC) mentioned that he was not allowed to cross the street outside his apartment complex without a parent, because 'it is too busy'. Children's limited licence to roam without adult supervision reveals clear implications for opportunities to make use of the diverse wellbeing affordances the central city offers (e.g. climbing a tree in Albert Park, viewing the changing sea at the harbour). The exploration, utilisation and transformation of wellbeing affordances fostered through the development of socio-spatial literacy and a positive sense of place at the larger neighbourhood level is often limited to places frequently visited with parents, which were also often chosen by them. The few quasi-natural environments within the central city did not appear to be sites that the children from Auckland Central wanted to show to the Beach Haven children as good places to play. This may be because they are relatively inaccessible because children have to rely on parents to access and spend time in these places (see also Ergler *et al.,* 2013). For example, during the Central City walk, Tim (CC) wanted to show Josh and Michael (BH) the 'sea' at Viaduct Harbour (a pedestrian promenade alongside moored yachts anchored adjacent to medium density apartment blocks) because he wanted to impress them with the great view, but then decided it was too far to walk, and took them to the newly rebuilt square around St Patrick's Cathedral instead. Within this square there are water features and some manicured green spaces, which are possibly regarded as 'stand-ins' for natural bush environments. However, the other central city participants did not show the Beach Haven children any parks in the area during the walk; the wellbeing affordances of the local environments differed between both locations. Taesong, for example, took his Beach Haven peers to a 'virtual playground' (game arcade) within a cinema and entertainment complex and showed them the indoor activities offered at Youthtown (an after school care facility) for 'fun and active things to do' (shown in Figure 13.2). It became obvious that Taesong (CC) spends much of his time in Youthtown, which offers a gathering place for children and supervised entertainment when parents are at work.

Similarly, McBeth (CC) was keen to show Dexter (BH) various shops within Auckland Central. The boys then spent a considerable amount of time in an electronics store playing on the gaming consoles, with McBeth (CC) explaining that these are good play spaces because they can play new videogames for free. Similarly, Taesong (CC) took his group to a local music shop where he regularly played with the instruments under the watchful eyes of staff. This anecdote speaks to McBeth (CC) and Taesong (CC) appropriating a space that has not been created for children per se to actualise wellbeing affordances as, for example, implied in a playground or park. City children appeared to feel at home within these stores, making use of their diverse wellbeing affordances (e.g. diving into and imagining virtual adventures, enjoying and creating their own soundscapes; see also Andrews *et al.*, 2014). A critical observation – as with Josh and Michael (BH) going into the Beach Haven bush alone – is the lack of parental presence. Here, a retail rather than 'natural' landscape is providing a setting for environmental literacy (in this case their ability to 'read' permission to try out games or music instruments and not be evicted). Although arguably generative of less physical activity and physical environmental knowledge, for city children this nonetheless aids a positive sense of self and attachment to neighbourhood places. The restorative wellbeing affordances therefore differ between locations.

Michael's (BH) fondness for outdoor play and connection to natural environments, which was evident during the walk in his home neighbourhood, was also shown during the walk in Auckland Central. For example, Michael (BH) quickly tired of an

Figure 13.3 Photograph showing impressions of an interactive outdoor exhibition.

Figure 13.4 Exploring the climbing trees during the central city walk.

interactive exhibition in a city square after they had played with the displays (see Figure 13.3). Michael's (BH) body language changed and it was evident that he was relieved when Auckland Central participant Tim (CC) led him into Albert Park. Once there, Michael ran excitedly and climbed the first tree he saw (see Figure 13.4). It seemed that the park felt familiar to Michael and reminded him of his own neighbourhood. This anecdote illustrates that through reading the environment (literacy) and activating a sense of place (a temporary at-home-ness among and within the trees), the familiar wellbeing affordances may be carried over and re*placed*.

Conclusion

In this chapter, we have outlined our conceptualisation of wellbeing affordances (comprised of a sense of place, place attachment and children's spatial literacy of social and physical environments) to highlight the complexity of urban children's wellbeing. We have argued that children's positive experiences of natural and mundane everyday city spaces afford opportunities for socio-spatial literacy and a strong sense of place that manifest themselves in numerous place-based wellbeing affordances (e.g. utilising the restorative or physical activity enhancing features of neighbourhood destinations). On a more general level, we have shown that becoming confident users of a range of urban settings has profound implications for feeling well, happy (see Chapter 6) and connected in a neighbourhood and, in turn, the formation of a strong sense of place.

We conclude that urban children's wellbeing is comprised of more than their physical or mental health – which dominates so many debates (e.g. see Chapters 2 and 3). Rather than partitioning types of health, we see benefit in embracing a wider reach implied by 'wellbeing' (Fleuret and Atkinson, 2007; Kearns and Andrews, 2010; Schwanen and Atkinson, 2015). This embrace requires a place-based approach that allows children's own experiences and conceptualisations of local wellbeing affordances to be heard. This hearing of experiences may be as subtle as some of the foregoing examples (e.g. collecting honeysuckles, playing music in a shop).

We have drawn on selected anecdotes from a novel process of photo-elicitation and neighbourhood walks in which local children 'hosted' others from a paired and profoundly different residential setting in the same city. Many of these anecdotes offered insights into the diversity of wellbeing affordances that different physical neighbourhood environments offer and the feelings they elicit. McBeth's (CC) story, for instance, indicated that place experience builds particular environmental knowledge that informs sense of place, feelings of belonging and self-identity. For him as a central city child, this was an identity that found some 'natural' aspects of Beach Haven, such as mud, as foreign and abhorrent. Beach Haven children, for instance, demonstrated clear feelings of attachment, belonging to, and care for, their close-to-home natural environments, while the Auckland Central children showed a sense of place within a near exclusively built environment. This observation is evident in some children's love of shops and their use of them as de facto playgrounds to gain similar feelings of belonging and connection that suburban children sought in wild as well as manicured natural environments. We also noted a transferability of environmental literacy, exemplified by Michael from Beach Haven who clearly 'read' the possibilities of mature trees in central Auckland's Albert Park and speculatively experienced at least an ephemeral sense of place away from his home place. This reveals the potential for the transferability of children's wellbeing affordances across a range of settings, but also runs the risk of learned affinities for, and actualisation of, familiar wellbeing affordances that are locally normalised.

Children's individual wellbeing and the health of the environment are entangled on many different scales. For example, our findings lend support to our contention that for a child to become socio-spatially aware of their neighbourhood setting, they need to actively explore their local environment and make use of the diverse wellbeing affordances; they need to develop a sense of place mentally and literally no matter whether this appears in a more natural or built-up setting. We can speculate that 'graduating' from adult accompaniment accelerates and cements the 'reading' of environments and eases the *re*placement of a 'sense of place' for their potential for play and active exploration. Autonomy, as well as relationships with the social, cultural and physical surroundings, is a significant dimension of wellbeing (Kearns *et al.*, 2006). Not only are the more apparent aspects of individual health (such as physical activity) supported by having the parental licence to explore, but so too is the development of personal identity and place attachment. This in turn fosters the maintenance and creation of healthy (green) spaces and contributes to planetary wellbeing. In other words, a recursive relationship between environmental literacy, a feeling of wellbeing and being well attached to neighbourhood places is demonstrated both on the individual and global scale.

In conclusion, because not all children have the benefit of green and 'natural' environments within their home neighbourhoods, there is a need to plan for and facilitate child-centred wellbeing affordances across a range of natural and built environments in order to foster opportunities for children to develop positive environmental experiences that build wellbeing. We therefore advocate for a two-way change. First, central city spaces should be retrofitted with an age-friendly design that affords all citizens engagement with, and provides choice within, a diverse range of 'wild' and 'manicured' nature spaces, as well as aesthetically and sensory-pleasing public spaces. Second, on the societal level a discursive shift is required to move away from the traditional 'child appropriate' play spaces towards a more flexible and creative engagement with the affordances cities can offer to enhance the wellbeing of children.

Notes

1 Children chose their own pseudonyms. The location the children are from will be indicated through an acronym reflecting their respective neighbourhood: BH for Beach Haven and CC for central city.
2 The park has 'wilderness' areas, access to the estuary and formal sports fields and a playground, as well as public outdoor exercise machines.

References

Andrews, G. J., Kingsbury, P. and Kearns, R. A. eds. 2014. *Soundscapes of wellbeing in popular music*. Farnham: Ashgate.

Bartos, A. E. 2013. Children sensing place. *Emotion, Space and Society*, 9, pp. 89–98.

Berkowitz, A. R., Archies, M. and Simmons, D. 1997. Defining environmental literacy: a call for action. *Bulletin of the Ecological Society of America*, 78, pp. 170–2.

Blaut, J. M. 1987. Place perception in perspective. *Journal of Environmental Psychology*, 7, pp. 297–305.

Carroll, P., Witen, K. and Kearns, R. A. 2011. Housing intensification in Auckland, New Zealand: implications for children and families. *Housing Studies*, 26, pp. 353–67.

Chawla, L. 1992. Childhood place attachment. In: I. Altman and S. Low, eds. *Place Attachment*. New York: Plenum Press, pp. 63–86.

Cresswell, T. 2014. *Place: a short introduction*. Malden, MA: Blackwell.

Davis, J. and Elliott, S., eds. 2014. *International Research in Early Childhood Education for Sustainability*. London: Routledge.

Dowling, R. 2000. Cultures of mothering and car use in suburban Sydney: a preliminary investigation. *Geoforum*, 31, pp. 345–53.

Duff, C. 2011. Networks, resources and agencies: on the character and production of enabling places. *Health & Place*, 17, pp. 149–56.

Ergler, C. R. 2011. Beyond passive participation: children as collaborators in understanding neighbourhood experience. *Graduate Journal of Asia–Pacific Studies*, 7, pp. 78–98.

Ergler, C. R. and Wood, B. E. 2015. Re-imagining youth participation in the 21st century: young people in Aotearoa New Zealand speak out. In: P. Kelly and A. Kamp, eds. *A critical youth studies for the 21st century*. Leiden: Brill.

Ergler, C. R., Kearns, R. A. and Witten, K. 2013. Seasonal and locational variations in children's play: implications for wellbeing. *Social Science & Medicine*, 91, pp. 178–85.

Eyles, J. and Williams, A., eds. 2008. *Sense of place, health and quality of life*. Aldershot: Ashgate.

Faber Taylor, A., Kuo, F. E. and Sullivan, W. C. 2002. Views of nature and self-discipline: evidence from inner city children. *Journal of Environmental Psychology*, 22, pp. 49–63.

Fleuret, S. and Atkinson, S. 2007. Wellbeing, health and geography: a critical review and research agenda. *New Zealand Geographer*, 63, pp. 106–18.

Freeman, C. and Tranter, P. J. 2011. *Children and their urban environment: changing worlds*. London: Earthscan.

Freeman, C. and Vass, E. 2010. Planning, maps, and children's lives: a cautionary tale. *Planning Theory & Practice*, 11, pp. 65–88.

Freeman, C., Van Heezik, Y., Hand, K. and Stein, A. 2015. Making cities more child- and nature-friendly: a child-focused study of nature connectedness in New Zealand cities. *Children, Youth and Environments*, 25, pp. 176–207.

Gibson, J. J. 1979. *The ecological approach to visual perception*. Boston, MA: Houghton Mifflin.

Gill, T. 2011. *Children and nature: a quasi-systematic review of the empirical evidence*. London: London Sustainable Development Commission.

Gill, T. 2014. The benefits of children's engagement with nature: a systematic literature review. *Children, Youth and Environments*, 24, pp. 10–34.

Gleeson, B. and Sipe, N. G. 2006. *Creating child-friendly cities: reinstating kids in the city*. London: Routledge.

202 Ergler, Kearns, de Melo and Coleman

Harris, P., Brown, B. and Wener, C. 1996. Privacy regulation and place attachment: predicting attachments to a student family housing facility. *Journal of Environmental Psychology*, 16, pp. 287–301.

Hart, R. 1979. *Children's experience of place*. New York: Irvington.

Heft, H. 1988. Affordances of children's environments. *Children's Environments Quarterly*, 5, pp. 29–37.

Hernandez, B., Hidalgo, M. C., Salazar-Laplace, M. E. and Hess, S. 2007. Place attachment and place identity in natives and non-natives. *Journal of Environmental Psychology*, 27, pp. 310–19.

Hidalgo, M. C. and Hernandez, B. 2001. Place attachment: conceptual and empirical questions. *Journal of Environmental Psychology*, 21, pp. 273–81.

Kahn, P. H. 2002. Children's affiliations with nature: structure, development and the problem of environmental generational amnesia. In: P. H. Kahn and S. Kellert, eds. *Children and nature: psychological, socio-cultural, and evolutionary investigations.* Cambridge, MA: MIT Press, pp. 93–116.

Kearns, R. and Andrews, G. J. 2010. Wellbeing. In: S. Smith, R. Pain, S. A. Marston and J. P. Jones, eds. *Handbook of Social Geographies*. 3 edn. London: SAGE Publications, pp. 309–28.

Kearns, R., McCreanor, T. and Witten, K. 2006. Healthy communities. In: M. Thompson-Fawcett and C. Freeman, eds. *Living together: towards inclusive settlements in New Zealand*. Melbourne: Oxford University Press, pp. 241–57.

Kytta, M. 2002. Affordances of children's environments in the context of cities, small towns, suburbs and rural villages in Finland and Belarus. *Journal of Environmental Psychology*, 22, pp. 109–23.

Lim, M. and Calabrese Barton, A. 2010. Exploring insideness in urban children's sense of place. *Journal of Environmental Psychology*, 30, pp. 328–37.

Loebach, J. and Gilliland, J. 2010. Child-led tours to uncover children's perceptions and use of neighbourhood environments. *Children, Youth and Environments*, 20, pp. 52–90.

Louv, R. 2005. *Last child in the woods: saving our children from nature-deficit disorder.* Chapel Hill, NC: Algonquin Books.

Malone, K. 2007. The bubble-wrap generation: children growing up in walled gardens. *Environmental Education Research*, 13, pp. 513–27.

McCreanor, T., Penney, L., Jensen, V., Witten, K., Kearns, R. and Moewaka Barnes, H. 2006. 'This is like my comfort zone': senses of place and belonging within Oruamo/ Beachhaven, New Zealand. *New Zealand Geographer*, 62, pp. 196–207.

McNeill, K. L. and Vaughn, M. H. 2012. Urban high school students' critical science agency: conceptual understandings and environmental actions around climate change. *Research in Science Education*, 42, pp. 373–99.

Mergen, B. 2003. Children and nature in history. *Environmental History*, 8(4), 643–69.

Moore, R. C. 1986. *Childhood's domain: play and places in child development*. London: Croom Helm.

Morris, J. L. and Zidenberg-Cherr, S. 2002. Garden-enhanced nutrition curriculum improves fourth-grade school children's knowledge of nutrition and preferences for some vegetables. *Journal of American Dietary Association*, 102, pp. 91–93.

Parkes, M. W., Morrison, K. E., Bunch, M. J., Hallstrom, L. K., Neudorffer, R. C., Venema, H. D. and Waltner-Toews, D. 2010. Towards integrated governance for water, health and social–ecological systems: the watershed governance prism. *Global Environmental Change*, 20, pp. 693–704.

Parkes, M., De Leeuw, S. and Greenwood, M. 2011. Warming up to the embodied context of First Nations child health: a critical intervention into and analysis of health and climate change research. *International Public Health Journal*, 2, pp. 477–85.

Relph, E. C. 1976. *Place and placelessness*. London: Pion.

Roth, C. 1992. *Environmental literacy: its roots, evolutions and directions in the 1990s*. Washington, DC: ERIC/CSMEE Ohio State University.

Schwanen, T. and Atkinson, S. 2015. Geographies of wellbeing: an introduction. *The Geographical Journal*, 181, pp. 98–101.

Spencer, C. P. and Blades, M., eds. 2006. *Children and their environments: learning, using and designing spaces*. Cambridge: Cambridge University Press.

Statistics New Zealand. 2013. *2013 Census tables about a place: Auckland region*. Available at www.stats.govt.nz/Census/2013-census/data-tables/tables-about-a-place.aspx?request_value=24394&tabname= (accessed 23 February 2017).

Timperio, A., Crawford, D., Telford, A. and Salmon, J. 2004. Perceptions about the local neighborhood and walking and cycling among children. *Preventive Medicine*, 38, pp. 39–47.

Tuan, Y. F. 1974. *Topophilia*. New York: Columbia University Press.

Tuan, Y. F. 1977. *Space and place: the perspective of experience*. London: Edward Arnold.

Wells, N. M. and Evans, G. W. 2003. Nearby nature: a buffer of life stress among rural children. *Environment and Behavior*, 35, pp. 311–30.

Wells, N. M. and Lekies, K. 2006. Nature and the life course: pathways from childhood nature experiences to adulthood environmentalism. *Children, Youth and Environments*, 16, pp. 1–24.

Williams, A., ed. 2007. *Therapeutic landscapes*. Aldershot: Ashgate.

Wilson, R. 1997. A sense of place. *Early Childhood Education Journal*, 24(3): 191–5.

Part IV
Viewing wellbeing

14 Mobilities of wellbeing in children's health promotion

Confronting urban settings in geographically informed theory and practice

Jeffrey R. Masuda, Paivi K. Abernethy, Linor David and Diana Lewis

In this chapter, we demonstrate the importance of thinking and practicing *beyond place*, in supporting the wellbeing of children living in cities. Just as 'health' has been widely critiqued as a concept too fixed into individual bodies and institutional settings, so too 'place' has come under increased scrutiny in recent years as being too metaphysically rigid to account for the highly dynamic – and mobile – complexity of urban lives. Hence in advocating for a focus on mobility and wellbeing, our approach is to blend recent arguments from the paradigm shift towards mobilities in human geography with our own observations as health promotion scholars and practitioners working in the field of children's health promotion in Canadian cities and neighbourhoods.

We set the stage with an introduction to the settings approach to health promotion, highlighting how the concept of settings has been widely misunderstood within health promotion. Second, we provide a brief outline of the 'new mobilities paradigm' to demonstrate why movement matters to people's wellbeing. An approach to geography that is attuned to the in-between spaces of people's lives has much to offer in revealing the geographies that can both facilitate and impede access to the city and its resources. Third, we introduce two practice insights from our observations in working in health promotion contexts with Indigenous and new immigrant populations, highlighting both the challenges that come with an under-appreciation of the mobilities of people's lives, as well as the opportunities of a mobilities perspective for improved practice. Lastly, we situate the mobilities of wellbeing within a normative theoretical position that espouses children's right to the city, suggesting that it is the city itself that needs to be positioned more centrally in health promotion theory and practice. The right to the city from the perspective of children and adults demands that personal and collective geographies, both in place and on the move, are accounted for equitably within urban life.

In order to establish a context for our chapter, we point the reader to Box 14.1, where author Linor David describes the scene of an unconventional health promotion setting that she recently found herself in.

Box 14.1 Encountering health promotion in a nail salon

For over a decade I've worked in a health promotion role in the city of Toronto, often with women who are new immigrants to Canada who are either pregnant or parenting. In 2012, while visiting a local nail salon as a customer, I had an encounter that changed my perspective about what my role was as a health promotion professional and, crucially, where my work belonged. It happened when I started chatting with the nail technician that I had come to know over the years. She was a young Vietnamese woman with a daughter the same age as my son. As it was summer, her daughter had accompanied her to the salon and was passing the afternoon watching TV. Ever the health promoter, I started to talk about low-cost camps that were available in the neighbourhood that could provide her daughter with better opportunities for physical activity and social interaction than at her mother's workplace. I thought that the camp would be a more suitable place for her health than the seemingly mundane setting of the salon. Her mother told me that it would be good for her to go there, citing the headaches and dizziness her daughter experiences from being at the salon all day. I was taken aback – this child was being made ill by breathing in the toxins at her mother's workplace – a space that for many women was paradoxically associated with self-care. The work that I had been doing for new immigrants on promoting safe and active lifestyles in the city would be of little help to the unique situation of this mother and her child. I wondered about the circumstances of other families that fell outside the spaces of my professional gaze. I realized that I need to see this child not just as a low-income neighbourhood resident and new immigrant child who might turn up at my community health centre seeking services, but as a social person with a body, family and community that is very much 'on the move'. Her health was very much contingent on her position and movement through a highly dynamic set of circumstances, including transnational labour mobility, cultural constructions of (nail) beauty, workplace environmental health regulation, and other social, economic and political circumstances.

'Setting' the stage

This illustration offers a starting point for our main focus in this chapter, which is to address what we see as a problematic impetus in health promotion to 'put children in their place' in both research and practice-based prescriptions for children's health and wellbeing. For many years, a socioecological emphasis in health promotion has inspired a 'settings' or, more broadly, 'place'-based episte-mology of health (Poland *et al.*, 1999; Soubhi and Potvin, 1999; Whitelaw *et al.*, 2001), offering a lens through which we are meant to understand and intervene in health beyond health care and to emphasise structural, contextual and community-based determinants of health. Inspired by the Ottawa Charter's (1986) proclamation that 'Health is created and lived by people within the settings of their everyday life; where they learn, work, play, and love' (World Health Organization, 1986), a specific healthy settings concept developed after the initial successes of the Healthy Cities Movement that began that same year. The World Health Organization (1998: 19) defines healthy settings as 'The place or social context in which people engage in daily activities in which environmental, organizational, and personal factors interact to affect health and wellbeing'.

In the nearly 30 years since its inception, both the academic literature and health promotion programmes are now rife with examples of settings-based approaches to health promotion to the point that the conceptual focus on 'settings' is implicit within programmatic foci such as healthy schools, healthy play spaces, healthy homes and even healthy hospitals (as more than clinical sites). These settings are often cited as key intervention opportunities for health promotion policy and practice across a wide variety of sectors, including public health, education, planning and research.

Of course, many positive things can be said of the settings approach to health promotion. First, a view to settings inspires us to work with people where they tend to live, work, play or especially where they seek help. This is particularly the case among low-income or other socially marginalised populations who tend to access more primary health care and social support services in community settings. Second, a settings approach shifts emphasis away from an individualist and behavioural to a more structuralist and socioecological epistemology of health and wellbeing. Third, settings, especially urban ones, bring together a wide array of stakeholders whose interests and activities intersect in particular institutional and geographical settings. Fourth, a focus on settings can improve the efficacy and efficiency of health promotion investments, particularly under conditions of time and/or resource scarcity. For those interested in addressing health inequities in particular, health promotion can specifically draw attention to adverse environmental or social conditions in housing, schools, neighbourhoods and perhaps even whole cities to inform resource distribution decisions that target more vulnerable populations and places.

Although recognising the importance of such strategic decisions, we also wish to highlight some of the possibly unanticipated consequences of the settings approach to children's health promotion, particularly when it is operationalised within the framework of existing institutional settings of health promotion practice and in professional epistemologies of health. Often, such settings are delineated at least in part on the basis of targeting priority populations – schools, neighbourhoods or clinics in minority or lower income neighbourhoods, often justified by higher incidences of chronic disease, lower levels of health literacy, and/or more propensity for health-averse behaviours. Crucially, as academics and practitioners of socioecological health promotion, we observe that a certain institutional fix has taken hold, which has co-opted much of the socioecological epistemology into a rhetoric for delivering individualist lifestyle-focused health promotion 'in the community'. We will refer throughout this chapter to this phenomenon as 'settings-as-institutions'. Such a co-optation, the substituting of an institutional setting, very much short-changes the ecological emphasis, which sees health and wellbeing as a function of a highly dynamic human–environmental web of relations.

Drawing on several epistemological axioms from our disciplinary associations as geographers, we identify at least five specific shortcomings of the settings-as-institutions approach in terms of its ability to understand health, place and in particular mobility, in framing health promotion interventions.

1   Settings-as-institutions often subordinates considerations about how populations end up in settings targeted for intervention. Instead, clients are presented as they appear 'in the problematic present' rather than as whole – historically and geographically constituted – people with complex life courses and geographies.

2   Settings-as-institutions fails to recognise the dialectic nature between people and place, leaving out the critical question of what happens to the settings themselves once targeted for intervention and improvement. There may be unintended consequences of well-intentioned health promotion interventions in settings, for example when concerns about safety and security turn play spaces into sites of surveillance that are antithetical to youth culture, creating a sense that adults find them untrustworthy (Fine *et al.*, 2003), or when neighbourhood improvement initiatives, ostensibly motivated by aspirations to improve the conditions for health, lead to gentrification and displacement of the very populations meant to be served (Gould and Lewis, 2012).

3   Settings-as-institutions does not account for the crucial role of agency in children's health, which may be denied as more and more everyday spaces – schoolyards, playgrounds and even whole neighbourhoods – are brought into the realm of professionalised intervention and under the control of adults.

4   Settings-as-institutions, particularly in urban contexts is often justified by quantitatively deduced sociodemographic constructs. Intervention sites are defined or justified by a matrix of census-based proxies of health determinants (e.g. minority areas or low-income quintiles) and poor health outcomes (e.g. incidence of low birth weights or chronic disease). For those whose bodies, identities and health are rendered visible by such associations, this approach possibly undermines their right to self-determination, including their right not to be labelled or included in state-sanctioned interventions.

5   We point to the notion introduced earlier that people are 'on the move' to illustrate the inherent lacuna within a settings-based approach to health promotion. Specifically, a focus on settings: (1) conflates a contextual epistemology of place with the institutional fix, theorising health promotion on the basis of where people, including children, 'ought to be' rather than where they actually are, which of course could be anywhere; and (2) under-emphasises the importance of being in, and moving through, the myriad *in-between* spaces and places of children's lives – between the home and the school, the neighbourhood and beyond, and often excluding sites where they can or even should be surveilled.

In the next section, we present an epistemological corrective called the new mobilities paradigm, which has been embraced by geographers over the past two decades, to point to ways in which health promotion can overcome the institutional fix inherent in the settings-based approach.

The new mobilities paradigm

For some time, a focus on mobility has taken hold among geographers out of a growing rejection of the disciplinary 'metaphysics of sedentarism' (Cresswell, 2001, 2010, 2012; Crang, 2002). Malkki (1992), an anthropologist, first indicted geography for its failure to theoretically account for 'movement' in its focus on theorising key geographic constructs such as places, territories and nations. She suggests that the very notion of 'rootedness' inherent in geographical approaches acquiesces to the current world order of global colonialism, nationhood and industrial capitalism. Mobility in conventional geographic thought is accused of being treated as anomaly, or a less theoretically important means, to a more theoretically important end in place and sense of place. But a plethora of research in the past decade or more has embraced the movement of people or ideas as a natural, vital and integrated part of human existence, replacing an epistemology of 'roots' with one of 'routes' (Gustafson, 2001). This turn to mobilities, or the 'new mobilities paradigm', (Sheller and Urry, 2006) has enabled attention to be drawn to issues, patterns, infrastructure and policies that traditionally tend to be treated as incongruent. As Cresswell (2012: 651) points out, mobilities research has been particularly beneficial in investigating 'interactions between ideas, people and things as they move and take hold (or fail to take hold)'. Elliott and Urry (2010: 7) have indicated that the new mobilities paradigm plays an increasing role in 'contemporary identity formation and re-formation'.

Cresswell (2010) identifies three constellations of mobility: movement, representation and practice. The physical act of movement merely involves getting from A to B, whether in the city, for example from the home to the grocer, church or resource centre, and to and from the city, for example from an Indian reserve (reserved lands set aside for the exclusive use of Canada's Indigenous people, which include First Nations, Métis or Inuit) or between countries. The relevance of such movement may involve the self, or a significant other whose mobility could have a bearing on health. Consider, for example, the impacts of children left behind when Filipino domestic workers come to Canada to find employment in order to support their families (Cortes, 2015). From this example, it is easy to imagine how mobility can also encompass things other than people, such as the circulation of money sent back home, the information of foreign recruitment agencies, or the goods, services and upwardly mobile socio-economic opportunities that may become available to families due to the work done on their behalf on the other side of the planet (see also Pratt, 2012; Clemens *et al.*, 2014).

Representation of movement refers to the shared meanings that are associated with particular mobilities of people, things or ideas and these too may be seen as healthful or harmful. Consider, for example, the notion of reserve-to-urban 'churn' (Norris and Clatworthy, 2011), a concept used to characterise the mobility pattern among Indigenous populations in Canada whose lives are, for better or worse, typically far more mobile than the non-Indigenous population. Indigenous peoples in Canada have become increasingly urbanised, with 54 per cent of the 1.17 million people who identify as Aboriginal living in urban centres (Snyder and Wilson, 2012) and residing disproportionately within lower-income inner city areas. Framing the mobility between and within reserves or cities as 'churn' implies an instability and itinerancy inherent among a people who it is assumed cannot achieve stability in their lives for lack of economic opportunity on the reserve and social support in the city. Such representations can have real impacts when children are subjected to institutional settings that consider such family mobilities as inconsistent with preconceived notions of safety and healthy development. The problem of treating urban Indigenous communities as homogenous and one-dimensional clusters defined by, and content to remain within, racialised spatial boundaries has long been recognised.

Practices of movement can have enormous influences on wellbeing. Such practices may be voluntary or coerced and may be accompanied by feelings of safety or joy when voluntary or, in the case of the suburbanisation of the poor, with social isolation, particularly when compounded with deficient public transit systems. Practices can also involve real and direct hardship, such as the common sight of young Indigenous mothers pushing strollers through deep snow alongside busy commuting corridors in Winnipeg's north end, only to have to wait at one of many unsheltered bus stops en route to whatever destination they are seeking to get to. Practices may also be influenced by, and have indirect health consequences when associated with, real and perceived encounters with racism emanating from the gaze of people 'on the way'. For instance, First Nations youth in Winnipeg have reported feeling a sense of foreboding or of feeling out of place in some commercial/retail spaces of the city (Skinner and Masuda, 2013). The prejudices of taxi and bus drivers, security staff and curious onlookers may contribute to making movement more difficult and undignified for Indigenous people of all ages and may concomitantly influence their movement decisions and thus limit their opportunities.

In drawing out these illustrations, we wish to highlight the likely significance of mobility in understanding 'progressive' – that is, socioecological – health promotion practice, focusing on practitioner perspectives of children's health promotion in urban settings. In doing so, we challenge the very premise of the setting as an exclusive or even appropriate lens through which to address health and wellbeing of children. We call for a conceptualisation of health promotion practice that recognises children in all of their (mobile) complexity; works to support children 'on the move' not just 'where they end up'; engages with the politics of movement to identity; and intervenes in the most health-disruptive mobilities. We take these ideas to critique mainstream approaches to settings-based health promotion through an exploration into the sedentary stasis that we have observed within interventions in our own professional activities.

The literature on children's wellbeing sees risk taking, exploration and discovery – or in other words, their physical and social 'movement' within the city – as crucial for the healthy social development of children, fostering autonomy, spatial literacy and geographical imagination (Freeman and Tranter, 2011; Malone, 2007). Taking examples from health promotion approaches for Indigenous people and recent immigrants to Canada as case illustrations, we question the premise of privileging the urban settings and even the urban itself in health promotion for these groups. Furthermore, in combining our academic and practitioner insights, we will propose a new concept 'mobilities of wellbeing' as an alternative approach to understanding complexities related to children's health promotion, which we argue can better address fundamental health inequities. Finally, in learning from our practitioner perspectives in light of the recent scholarship on the geography of mobility we hope to chart a more fluid and importantly agentic model of children's health promotion that prioritises mobilities of wellbeing as fundamental to their 'right to health in the city'.

Presenting the practice challenge

As health promotion 'academic practitioners' (that is, professionals who have long experiences in working 'on the ground' alongside personal pursuits of higher education and research) whose experience has been largely concerned with the health and wellbeing of minorities, we realise that people over their life courses and in their everyday lives experience fluctuating multicultural identities and face ongoing intercultural negotiations that are unique to the contemporary era of global colonialism generally and in its many manifestations in Canada particularly. For instance, when Indigenous people and immigrants began to move into cities over the last century, they retained contact with their geographic origins. In the past, letters and occasional long and cumbersome journeys were required to maintain contact with friends and relatives, but such mobility has accelerated significantly along with the advent of modern communications technology and more rapid and (relatively) affordable transportation infrastructure, all to say that mobility is on the rise.

Despite this recognition, our work is often framed around constructs of population-based rigid structural and normative terms and in binaries such as rural or urban, Indigenous or mainstream, and white or visible minorities. These globalised urban dwellers, often presented to us as 'minority or underprivileged populations' who happen to cluster around our own professionalised locales, are indeed globally mobile subjects whose lives are both enriched and limited by the cultural interchange they maintain with distant social, institutional and cultural relations. Boxes 14.2 and 14.3 provide practice insights from the fields of Indigenous and new immigrant health promotion that highlight our critique of the rigid normative classifications and sedentarist labelling that we see as hindering our work. This practice background will give us insight into the rural–urban and local–global stasis inherent in our own work with Indigenous people and new immigrants.

Box 14.2 Practice insight: Indigenous health on the move

For Indigenous people, operationalising health promotion within an existing urban institutional setting departs from well-established health frameworks developed by Indigenous scholars and health experts over the years. For example, the National Aboriginal Health Organization (NAHO) provides a relational definition of health as the balance between the physical, mental, emotional and spiritual realms as well as the environment, culture, family and community, and that wellbeing flows from balance and harmony among all elements of personal and collective life (First Nations Centre at NAHO, 2007). The Integrated Life Course and Social Determinants Model of Aboriginal Health recognises that although social determinants are important to assess health in general, health disparities that exist between Indigenous and non-Indigenous people are compounded by colonisation, social exclusion, loss of language and culture, land displacement and environmental dis-possession (Reading and Wien, 2009). Traditional ties to the natural environment and cultural continuity are also important considerations that must be measured when assessing the health of Indigenous people (Adelson, 2005; King *et al.*, 2009).

Adhering to relational Indigenous epistemologies in their work, authors Lewis and David have learned to ask several key questions in their health promotion approach, including what gets carried with people (language, tradition, ancestry, food) when they move (understanding that mobility does not necessitate a loss of attachment to environment, culture, family and community); what gets taken away or left behind; how does mobility influence parenting and other environmental factors that influence child development; what is useful that is gained by moving; and maybe most importantly, how do health promotion practitioners address vulnerable areas (precarious environmental conditions related to child rearing) in a way that opens eyes and respects the whole person without instilling fear, imposing harmful normative assumptions, or taking away important aspects of cultural identity that people carry with them?

Lobo (1998) critiques the view of Indigenous ways of life as somehow being exclusive to notions of rurality and nature. She highlights the multi-faceted and diverse ways that Indigenous people perceive their environment 'based on a series of very dynamic relationships and shared meanings, history, and symbols, rather than based on the more commonly assumed clustered residential and commercial neighbourhood' (1998: 101). We have come to realise that the conceptual classification of 'the urban' in itself may be too rigid and oversimplified framing for health promotion issues, particularly in an era when most of the global population now lives in cities, and where the sphere of urban influence reaches to the farthest corners of the (rural) world.

Narrow generalisations such as 'urban health promotion' distort the focus and trivialise vital health determinants when problematising children's environmental health as an arena for health promotion intervention. In Indigenous children's health promotion, there is a long history, particularly in colonial states like Canada and Australia, of categorising families of a specific cultural background as 'at-risk' or 'vulnerable' that indirectly blames parents for being incapable of taking care of their families. Take, for instance, the increasingly accepted indictment of Canada's Residential Schools as an act of cultural genocide, meant to metaphorically and culturally 'kill the Indian in the child' (Harper, 2008) and often literally to commit Indigenous children to early deaths. Policies that attempt to fix people in place (reserves, derelict districts, service dependent communities) discursively remove them as functional and respected members of the society and treat them as less competent – ironically as a consequence of the very policies that marginalise them in the first place.

For instance, health statistics present the teen birth rate among Indigenous women as significantly higher (1 in 10) than the teen birth rate of non-Indigenous Canadian women (1 in 50) (Guèvremont and Kohen, 2013). Through statistics we are made to understand that children of teen mothers off-reserve (i.e. mainly in urban centres) are more likely to have dental problems than children of older mothers off-reserve, and have lower prosocial scores (sharing, getting along), higher emotional symptoms (worries, fears), higher inattention-hyperactivity and higher conduct problems (fighting, bullying) (Guèvremont and Kohen, 2013). However, these findings provide no recognition or guidance to understand the continuum of conflicting intercultural, social and environmental factors that influence these Indigenous parents and their children. The contemporary approach to target populations intrinsically ignores the ways in which, for instance, mobilities of both people and ideas influence the cultural identity and location of these young mothers, their wellbeing, parenting ability and consequently children's health.

Indigenous groups not only experience higher mobility than the non-Indigenous population, but the mobility patterns between Indigenous groups also vary, and may have important implications, for instance, in accessing health and social support services. Norris and Clatworthy (2003) identified how the differences in mobilities often depend on an individual's legal status and connections to reserve communities as certain rights and benefits accrue from that attachment, such as housing, education, tax exemptions and other services. Indeed, most Indigenous people living in urban areas maintain a sense of connection to their home community or place of origin, which is integral to strong family and social ties and culture (Environics Institute, 2010). The proximity of many First Nation and Métis communities to cities, the ancestral nomadic culture of many Indigenous groups such as the Innu (Neuwelt *et al.,* 1992), and the strong connection between traditional land, culture and wellbeing also enhance close connections to home community or place of origin for Indigenous urban dwellers.

The colonial frameworks that dictate the geographies of 'Aboriginal rights' further complicate the relationship between place and wellbeing because certain rights associated with reserves do not apply in urban settings. The term 'Aboriginal rights' refers to the inherent, collective rights of Indigenous people that are understood as sovereignty to govern the land, resources and one's own culture based on pre-colonial social orders and the original occupation of a particular region. Legally, however, this concept has been convoluted by the Treaties (between the Crown and various Indigenous peoples) as well as the Indian Act, the latter of which artificially determines who has the right to have official Indigenous identity, what the Indigenous rights include, and where or how they can be practiced, all of which are firmly rooted in notions of place and belonging that hinge on the reserve. In general, Indigenous people perceive this form of governance as disrespectful of a view of 'Aboriginal rights' associated with a less spatially bound way of life, particularly among those who live in urban settings. In cities, Indigenous people face discrimination and racism at all levels, from individual to structural, which consequently manifest in many psychosocial impacts, particularly among children. These experiences and impacts are only exacerbated by the stress that is associated with efforts to access Aboriginal rights while out of place (Senese and Wilson, 2013). The results are the consequent poor health outcomes, such as depression, anxiety or other health damaging behaviours that are ironically all-too-often used in population health reporting as the basis for statistically defining these populations in the first place.

The institutionalised culture of disrespect influences the development and wellbeing of urban Indigenous children. Indeed, Indigenous families experience significantly higher levels of perinatal and infant health challenges than the non-Indigenous Canadians (van Herk *et al.*, 2012; Guèvremont and Kohen, 2013). This includes low birth weight, higher infant mortality, gestational diabetes, foetal alcohol syndrome and teenage pregnancies. Preventive services such as prenatal education, breastfeeding support and preschool playgroups are known to mitigate these risks and are considered to be an important part of health interventions developed to reduce the health inequities that are reinforced by colonial relations. However, urban Indigenous families are less likely to access these services due to experiences of systemic discrimination and marginalisation within health care encounters, again often because they are seen to be out of place by service providers who harbour not only their own prejudices, but also professional standards of practice associated with white, middle-class and masculine culture. On the other hand, service providers who understand the social, familial, historical, political and geographic diversity and the associated legal challenges of Indigenous people living in urban centres can foster a sense of belonging by creating a safe place that provides access to larger social networks that reflect one's own cultural values and views of wellness (van Herk *et al.*, 2012). In our observations, supports that are provided to people within the context of their highly dynamic and tenuous urban lives remain the exception rather than the norm in most sites of health care and health promotion delivery in Canada.

Box 14.3 Practice insight: newcomers, goers and comebackers again

A similar illustration to that of Indigenous populations can be found in statistical representations of new immigrant populations, likewise often generated in population health statistics by categories such as ethnic origin. The very notion of ethnicity is related to the sedentary metaphysics of place, implying a rootedness in a 'land' (Malkki, 1992), which belies the contemporary experience of many new Canadians whose lives before and after coming to Canada can be very much unrooted, and often uprooted. Representations of ethnicity epitomise the problem of metaphysical sedentarism, conflating two independent and not necessarily related characteristics: social constructions of race (ethnic) that are based on geography (origin). Health statistics of ethnicity assume a unity between identity, culture and place that has less and less relevance in today's pluralist and globalised society. For instance, 'ethnic origin' does not account for the length of stay in country, including whether a person was born in Canada or not, the many different cultural encounters here and abroad that may influence one's upbringing or life course, or the impacts of other aspects of identity formation such as religion, popular culture, recreation or education. Ethnicity is likewise oblivious to the distinction between one's genetic composition and geographical lineage, which, although perhaps historically convergent, is not an interchangeable construct in defining people's place in society nor their health experiences. Indeed, beyond the implicit association with phenotype (e.g. 'ethnic people' have a skin colour other than 'white'), the value of this category from a health perspective is of declining importance. Such categorisation unintentionally creates another disruptive and unnecessary spatial alienation, assuming 'white' European origin as 'authentically' Canadian and expecting Canadians of several generations with skin tones other than white to highlight their ethnic origin as a means to benchmark their social distance from this norm.

There is a growing base of research that critiques the place of immigrant identities in health research (Ong, 2006; Mountz, 2011), binaries of 'good' and 'bad' immigrants (Razack, 1999; Hyndman, 2000), and public discourses about migration (Cresswell, 2006), but health promotion studies looking into dynamics of cultural identities, mobilities and health trajectories of immigrant populations are certainly less visible. With the exception of refugees, most new immigrants' explicit desire to change location to gain some socio-economic mobility is voluntary, at least in a political sense. But immigrants are often confronted with greater challenges than they expect, owing to the fact that social mobility is 'a resource to which not everyone has an equal relationship' (Sheller and Urry, 2004: 5). Cultural integration is a dynamic process that is increasingly complicated by greater rates of travel and new communications technologies that blur the lines between the local here and now and global origins and mobilities.

For example, when working with Chinese new immigrants in Toronto in multiple pregnancy and parenting programs over the past 10 years, author David noted how many parents deal with the challenge of high childcare costs. For many working or student parents the solution to this dilemma has often been to send their children back to China to live with grandparents or other relatives. Many women have been hesitant to talk about their situations, perceiving that professionals expect them to be the primary caregivers for their children and would judge them negatively. The issue of family separation and reunification can have long-term emotional ramifications for both parents and their children (Tate, 2011). Providing appropriate services has therefore required that David consider the mobilities of her clients' lives in her work to acknowledge the complicated nature of their global trajectories and the strain of settlement and tenuous socio-economic circumstances.

The sense of powerlessness that is created by the lack of access to socio-economic mobility as well as internal negotiations between one's place of origin and values, norms and practices within Canada can create a deep source of stress and erode the self-confidence of new immigrants (Walters *et al.*, 2007). In the context of children's health promotion, this creates another layer of complexity for practitioners working within existing institutional and discursive constraints. Questions arise as to how practitioners can bring attention to children's health issues when working with marginalised and vulnerable clients, knowing that most have the will to change their environment but lack the necessary resources to do so. Practitioners are placed in a paradoxical situation where efforts to promote their clients' children's health may be adding to existing stresses by outlining the risks of unhealthy housing, toxic workplaces or other unfriendly settings for healthy development.

The continuous process of internal and external cultural negotiation that immigrants experience was demonstrated in a parenting workshop series that David arranged for newcomer women to Toronto in 2013. In order to reflect upon parenting styles, all participants were given a lump of play-dough and asked to shape out of the clay the animal that most represented their parents' parenting style. The participating mothers produced a wide range of animals, ranging from bears to mice and ants. Many animals represented a parenting style that was either disciplinary or hard working, with an emphasis on sacrificing being 'present' for children in order to provide their basic needs. For the second part of the workshop, parents were asked to create the animal that they most desired their parenting style to reflect. In almost everyone's case, a new animal appeared that more closely represented the current North American attachment paradigm. For many, moving to Toronto meant adopting a parenting style that was different to what they had experienced outside of Canada. In the discussions following the exercise, the mothers gave various justifications for their desires to master new parenting models. In part, the change reflected their desire to fit into the new culture, but the mothers also recognised that they were learning new things about how children developed. Notably, the new parents wanted to differentiate themselves from their parents' generation. This example illustrates the need for a more nuanced view on the fluidity (and mobility) of cultural identity in health promotion practice, particularly within highly diverse urban contexts.

Concluding thoughts: combining 'mobilities of wellbeing' and children's right to health in the city

The new mobilities paradigm is not a panacea for the myriad structural challenges in addressing the health aspirations of children and their families. Instead, our purpose here is to emphasise a heretofore-neglected epistemology of health and wellbeing that may help to improve health promotion practice, including and especially among those practices that take a socioecological and settings-based perspective. We believe that the new mobilities paradigm in geography can direct health promotion research and practice towards new types of questions that need to be asked when identifying community needs and when developing effective and culturally appropriate interventions – that is, one that accounts for the fluidity of culture, identity and place as derivatives of both roots and routes, including the movement of people, places, ideas, health promotion resources and services, policies and regulations, and any other dimension of a highly mobile social life.

The greatest challenges in integrating mobility into health promotion practice are the existing momentum of disciplinary discourses, coupled with an easy reliance of researchers, policy makers and practitioners alike on the statistical representation of populations, problems and even solutions. Whereas much of the work on mobilities in geography is of an abstract, theoretical purpose (as useful as these may be in their own right), for health promotion, a normative turn is needed to put these ideas to work, particularly when encountering rigid institutional fixes often faced by practitioners on the ground. To embrace this difference in a meaningful manner, we point to the concept of 'mobilities of wellbeing' that shifts the metaphysics of 'health' away from one that fixes individual bodies and whole populations *within* places of health promotion intervention and more towards a more dynamic and fluid concept that accounts for people 'where they are' and 'on the move', whether such places and movements are voluntary or coerced, healthful or harmful.

At its heart, mobilities of wellbeing can bring the health promotion practitioner's attention to bear on the city itself, one that can take into account epistemologically and institutionally the complexity of people, places, things, ideas, resources and information that are very much in motion within the city and beyond. Such a view reminds the practitioner that although people do inhabit and occupy myriad settings on their daily journeys, they are not defined or constrained by such settings. Efforts to build a healthy school do not need to limit activities to the classroom, lunchroom or playground, but might extend into the community. Likewise, the designation of a healthy city does not end at the rural frontier, but extends relationally across the rural countryside and beyond provincial and national borders as children's lives and those of their families are rendered more mobile from the structural oppressions and displacements of global neoliberal urbanisation.

In principle, a sensitivity to mobility can be integrated into many existing health promotion frameworks, offering a more fluid conceptual model than the static yet firmly 'fixed' representations of health and wellbeing inherent in many current approaches. Furthermore, by taking seriously the critique that a more mobile metaphysical sensibility can offer, we might redefine the purposes, mechanisms and successes of health promotion and health equity work in ways that directly confront, rather than depend upon currently hegemonic institutional fixes in health and health care. We might instead see health promotion as one part of an effort to assert people's right to the city – that is, the right to live lives within, through and between places and social contexts that are healthful and not harmful, whether in or on the way to the clinic, community centre, school, grocer, playground, nail salon or reserve. In such a representation, health promotion can aim to be more than helping people (to help themselves) and instead seek to help the city itself change in a way that ensures that all children are given the opportunity to be accepted and supported no matter where they go and how they get there.

References

Adelson, N. 2005. The embodiment of inequity: health disparities in Aboriginal Canada. *Canadian Journal of Public Health/Revue Canadienne de Sante'e Publique*, 96, pp. S45–61.

Clemens, M. A., Ozden, C. and Rapoport, H. 2014. Migration and development research is moving far beyond remittances. *World Development*, 64, pp. 121–4.

Cortes, P. 2015. The feminization of international migration and its effects on the children left behind: evidence from the Philippines. *World Development*, 65, pp. 62–78.

Crang, M. 2002. Between places: producing hubs, flows, and networks. *Environment and Planning A*, 34(4), pp. 569–74.

Cresswell, T. 2001. Mobilities. *New Frontiers*, 43, pp. 4–28.

Cresswell, T. 2006. *On the move: mobility in the modern western world*. London: Routledge.

Cresswell, T. 2010. Towards a politics of mobility. *Environment and Planning D, Society and Space*, 28(1), pp. 17–31.

Cresswell, T. 2012. Mobilities II still. *Progress in Human Geography*, 36(5), pp. 645–53.

Elliott, A. and Urry, J. 2010. *Mobile lives*. New York: Routledge.

Environics Institute, 2010. *Urban Aboriginal Peoples Study: main report*. Available at www.uaps.ca/wp-content/uploads/2010/03/UAPS-Main-Report.pdf (accessed 28 February 2017).

Fine, M., Freudenberg, N., Payne, Y., Perkins, T., Smith, K. and Wanzer, K. 2003. 'Anything can happen with police around': urban youth evaluate strategies of surveillance in public places. *Journal of Social Issues*, 59(1), pp. 141–58.

First Nations Centre @ NAHO. 2007. *Understanding health indicators*. Available at www.naho.ca/documents/fnc/english/FNC-UnderstandingHealthIndicators_001.pdf.

Freeman, C. and Tranter, P. J. 2011. *Children and their urban environment: changing worlds*. London: Earthscan.

Gould, K. A. and Lewis, T. L. 2012. The environmental injustice of green gentrification. In: J. N. DeSena and T. Shortell, eds. *The world in Brooklyn: gentrification, immigration, and ethnic politics in a Global City*. Plymouth: Lexington Books. pp. 113–46.

Guèvremont, A. and Kohen, D. 2013. Do factors other than SES explain differences in child outcomes between children of teenage and older mothers for off-reserve First Nations Children?. *The International Indigenous Policy Journal*, 4(3), p. 6.

Gustafson, P. 2001. Roots and routes exploring the relationship between place attachment and mobility. *Environment and Behavior*, 33(5), pp. 667–86.

Harper, S. 2008. Statement of apology to former students of Indian residential schools. Government of Canada. Available at www.aadnc-aandc.gc.ca/eng/1100100015644/1100100015649 (accessed 28 February 2017).

Hyndman, J. 2000. *Managing displacement*. Minneapolis, MN: University of Minnesota Press.

King, M., Smith, A. and Gracey, M. 2009. Indigenous health part 2: the underlying causes of the health gap. *Lancet*, 374(9683), pp. 76–85.

Lobo, S. 1998. Is urban a person or a place? Characteristics of urban Indian country. *American Indian Culture & Research Journal*, 22(4), pp. 89–102.

Malkki, L. 1992. National geographic: the rooting of peoples and the territorialization of national identity among scholars and refugees. *Cultural Anthropology*, 7(1), pp. 24–44.

Malone, K. 2007. The bubble-wrap generation: children growing up in walled gardens. *Environmental Education Research*, 13, pp. 513–27.

Mountz, A. 2011. Refugees: performing distinction: paradoxical positionings of the displaced. In: T. Cresswell and P. Merriman eds., *Geographies of mobilities: practices, spaces, subjects*. Farnham: Ashgate. pp. 255–70.

Neuwelt, P. M., Kearns, R. A., Hunter, D. J. W. and Batten, J. 1992. Ethnicity, morbidity and health service utilisation in two Labrador communities. *Social Science and Medicine* 34, pp. 151–60.

Norris, M. J. and Clatworthy, S. 2003. Aboriginal mobility and migration within urban Canada: outcomes, factors and implications. In: D. Newhouse and E. Peters, eds. *Not strangers in these parts: Urban Aboriginal Peoples*, pp. 51–78.

Norris, M. J. and Clatworthy, S. 2011. Aboriginal mobility and migration within Canada's Friendship Centre Areas: patterns, levels and implications based on the 2006 Census. In: P. Dinsdale, J. White and C. Hanselmann, eds., *Urban Aboriginal communities in Canada: complexities, challenges, and opportunities*. Toronto, ON: Thompson Educational, pp. 173–218.

Ong, A. 2006. Mutations in citizenship. *Theory, Culture & Society*, 23(2–3), pp. 499–505.

Poland, B. D., Green, L. W. and Rootman, I., eds. 1999. *Settings for health promotion: linking theory and practice*. Thousand Oaks, CA: SAGE Publications.

Pratt, G. 2012. *Families apart: migrating mothers and the conflicts of labor and love.* Minneapolis, MN: University of Minnesota Press.

Razack, S. 1999. *Looking white people in the eye: gender, race and culture in courtrooms and classrooms.* Toronto, ON: University of Toronto Press.

Reading, C. and Wien, F. 2009. *Health inequalities and social determinants of Aboriginal peoples' health.* Available at www.nccah-ccnsa.ca/Publications/Lists/Publications/Attachments/46/health_inequalities_EN_web.pdf (accessed 28 February 2017).

Senese, L. C. and Wilson, K. 2013. Aboriginal urbanization and rights in Canada: examining implications for health. *Social Science & Medicine*, 91, pp. 219–28.

Sheller, M. and Urry, J. eds. 2004. *Tourism mobilities: places to play, places in play.* New York: Routledge.

Sheller, M. and Urry, J. 2006. The new mobilities paradigm. *Environment and Planning A*, 38, pp. 207–26.

Skinner, E. and Masuda, J. R. 2013. Right to a healthy city? Examining the relationship between urban space and health inequity by Aboriginal youth artist-activists in Winnipeg. *Social Science & Medicine*, 91, pp. 210–18.

Snyder, M. and Wilson, K. 2012. Urban Aboriginal mobility in Canada: examining the association with health care utilization. *Social Science & Medicine*, 75(12), pp. 2420–24.

Soubhi, H. and Potvin, L. 1999. Home and families as health promotion settings. In: B. D. Poland, L. W. Green and I. Rootman, eds. *Settings for health promotion: linking theory and practice.* Thousand Oaks, CA: SAGE Publications, pp. 44–85.

Tate, E. 2011. *Family separation and reunification of newcomers in Toronto: what does the literature say?* Available at http://www1.toronto.ca/city_of_toronto/toronto_public_health/healthy_communities/mental_health/family_separation/files/pdf/fsr_newcomers.pdf (accessed 28 February 2017).

van Herk, K. A., Smith, D. and Gold, S. T. 2012. Safe care spaces and places: exploring urban Aboriginal families' access to preventive care. *Health & Place*, 18(3), pp. 649–56.

Walters, D., Phythian, K. and Anisef, P. 2007. The acculturation of Canadian immigrants: determinants of ethnic identification with the host society. *Canadian Review of Sociology/ Revue Canadienne de Sociologie*, 44, pp. 37–64.

Whitelaw, S., Baxendale, A., Bryce, C., MacHardy, L., Young, I. and Witney, E. 2001. 'Settings' based health promotion: a review. *Health Promotion International*, 16(4), pp. 339–53.

World Health Organization (WHO). 1986. *The Ottawa Charter for Health Promotion.* First International Conference on Health Promotion, Ottawa, 21 November 1986. Geneva: WHO. Available at www.who.int/healthpromotion/conferences/previous/ottawa/en/ (accessed 28 February 2017).

World Health Organization (WHO). 1998. *Health promotion glossary.* Geneva: WHO.

15 Identity, place, and the (cultural) wellbeing of Indigenous children

Michelle Thompson-Fawcett and Robin Quigg

Holistic Indigenous understandings of wellbeing typically integrate health and development with the physical, spiritual and community environment where a people have their ancestral standing. Accordingly, Indigenous children's meaning in life and self-worth have a particular connection to the place of this historic belonging. Their actions and behaviours may reflect an ancestral integration with landforms and may be manifest in physical and spiritual practices. However, most research on Indigenous wellbeing has been founded on non-Indigenous notions of health rather than broader Indigenous conceptions of wellbeing (King *et al.*, 2009). This chapter, by contrast, explores the associations between the cultural wellbeing of Indigenous children and their identity in place. It does so in the colonial context. In other words, the primary focus is on places where imperialism and its patriarchal traditions have endorsed a refusal to respect and celebrate cultural difference.

In that context, we suggest that recovering and refreshing a traditional ethic of 'locatedness' would be highly beneficial in (and for) the twenty-first century. Such an ethic would embrace the 'particularities of land, of language, of history and of culture' (Martin, 2000: 82) and reawaken a focus on the holistic wellbeing of Indigenous children commensurate with (post)-colonial needs. We are not implying that Indigenous peoples are passive recipients of a colonial existence, but rather that although dominant colonial practices reproduce Western ways of being, in recent decades there has been a significant resurgence of resistance through the use of Indigenous languages, knowledges and culture. Our chapter therefore seeks to offer encouragement. The basis for optimism comes from the success of initiatives that reassert the potency and integrity of Indigenous philosophies and actions. These philosophies and actions challenge how broader society envisages the wellbeing of Indigenous children in colonised locations.

The reality of cultural heterogeneity in (post)-colonial societies begs the need for a 'politics that treats difference as variation and specificity rather than as exclusive opposition, aims for a society and polity where there is a social equality among explicitly differentiated groups who conceive of themselves as dwelling together without exclusions' (Young, 1995: 165). How might a shift in this direction be achieved? How might we honour variation and live positively alongside difference?

This chapter assembles ideas from Indigenous groups in response to these questions. Although the literature on Indigenous children's wellbeing – in place – is quite limited, key commonalities in relevant material are presented, primarily sourced from Canada, New Zealand, the US, Australia, Scandinavia and Africa. The chapter begins with a brief introduction to Indigenous health. We then set out some ways of understanding the wellbeing of Indigenous children. The two sections prior to the conclusion present the primary challenges we address: first the threats and barriers to Indigenous children's wellbeing, and then some fundamental principles for transforming the current situation. Although the chapter presents some general principles held in common across many Indigenous groups, we recognise that in practice, diverse geographic, cultural, historical and societal contexts – within and between Indigenous groups – mean that in any particular setting, a regional specificity of approach is crucial to achieving the wellbeing of Indigenous children. Hence, in the chapter we want to avoid implying that we are in any way essentialising or conflating Indigenous diversity.

Indigenous 'health'

The health of Indigenous people in many places in the world is poorer than non-Indigenous people (Robson and Harris, 2007; Borell *et al.*, 2009; Tobias *et al.*, 2009). In particular, the effects of colonisation are still felt acutely in Indigenous communities today, with alienation from ancestral lands and traditional customs, and introduction of both harmful substances, such as tobacco and alcohol, and diseases often combined with impoverished lifestyles in urban settings far from traditional home bases (Clark, 2009; Skinner and Masuda, 2013). Indigenous peoples' relationship with their environment, and their management of it so that it sustains them nutritionally and spiritually, has been marginalised if not severed completely. Although much of this colonisation may have occurred in the 1800s, its ongoing effects are still observed with Indigenous groups disproportionately disadvantaged in relation to health, education, income and crime in many countries (Peters, 2000; Gracey and King, 2009). Disenfranchisement from traditional locations, language, knowledge, culture and practices adds considerable complexity to meaningfully understanding contemporary health and wellbeing.

Children are vulnerable to the effects of poor health in their communities. In fact, they bear a high burden of the diseases and substandard outcomes, both in physical and mental terms. There are enduring effects from inadequate perinatal and neonatal care, as well as substandard living conditions, nutrition and exposure to acute and chronic diseases. These childhood conditions and situations disadvantage the children in later life (Gracey and King, 2009).

Improving the health of Indigenous people requires recognition of their whole person and their links to the environment and their communities, rather than a focus on their disease (King *et al.*, 2009). Reducing disparities requires acknowledgement that the health of Indigenous people is integrated within their worldview (Durie, 2004). In regard to children, there is a clear need to understand the aspirations of Indigenous groups for childhood cultural wellbeing and its identity in, and relationship with, ancestral locations and histories.

The wellbeing geographies of Indigenous children

The field of health science conjures up visions of sick people, and the health literature can easily focus on prevalence of diseases, incidence of diseases and discussions about co-morbidity and DALYs (disability-adjusted life year). The statistics and stories of illness and death rates for Indigenous children usually paint a poor picture – they tend to be less healthy, less educated and less wealthy than their non-indigenous peers, no matter where in the world they live (for example, Robson and Harris, 2007).

In Indigenous sciences, however, 'the world is often understood in terms of flows of energies (and sometimes entities) across a permeable boundary between manifest and unmanifest realities' (Herman, 2008: 75). In other words, spiritual and physical aspects of the world are important to health and wellbeing. The physical world is not merely a 'container for human life' (ibid.), 'geohistorical roots to an ethnic homeland' (ibid.) emphasise the integrated nature of mind and matter and the personal (ancestral) relationships with the natural world in a particular place. This establishes a strong 'connection between human health and the integrity of nature, and the rights of both. . . . Consequently we see ourselves as not only "of the land" but "as the land"' (Te Aho, 2011: 347). In this way, the spiritual and physical wellbeing of children is linked both directly and indirectly to the wellbeing of the environment. Sustaining the environment in a manner consistent with Indigenous values is important for the self-esteem and confidence of the community. Identity, place and wellbeing are deeply intertwined. Indigenous groups each have their own aspirations for their children in terms of culture, identity and relationship with ancestral places and any alienation from traditional land, language and culture as a result of colonisation can make this important connection very difficult to maintain – but often all the more important.

Issues of colonisation affecting wellbeing of Indigenous children

Understanding the impact on wellbeing of 'Indigenous-specific factors related to colonisation' is indispensable to devising pathways forward (King *et al.*, 2009: 76). This section sets out the breadth of complicating issues that are persistently reported in literature related to the cultural wellbeing of Indigenous children. The factors presented here are not exhaustive, but rather offer an overview of the problematic context with which Indigenous groups work as they seek to secure the wellbeing of their children.

The effects of colonisation on Indigenous cultures have been inestimable (for example, Borell *et al.*, 2009) and in subsequent decades there has been a wide range of efforts aimed at decolonising, reasserting cultural principles and systems, and achieving self-determination (Peters, 2000). It is little wonder, therefore, that contemporary Indigenous values in regard to child health and wellbeing have been based on notions of cultural survival and emancipatory practices.

A variety of imposed transitions over many decades have significantly impacted Indigenous peoples. The ongoing suppression, distortion and reconstruction of traditional cultures have placed such groups under relentless pressure. Changing identities, loss of resources, quashing of spiritual mainstays and demise of authority are major contributors to a diminished wellbeing. For example, for many Indigenous groups, kinship relationships are the primary key to authentic identity and notions of wellbeing. But dramatic changes resulting from colonisation have transformed the structure and operation of family, clan and homeland connections, diminishing a sense of belonging and leading to difficulty dealing with the conflict 'between contradictory value systems' (Kral *et al.*, 2011: 435). Indigenous groups have found themselves enmeshed in competing and contradictory worlds, particularly as a result of the advent of new value systems that counter those of their elders. A case in point has been that of the introduction of stereotypical Western gender roles that were 'not evident in pre-colonial society' (Smith, 2000: 58). In many Indigenous societies this has led to the replacing of more reciprocal male/female roles, introducing notions of patriarchy that have subsequently institutionalised colonial values privileging males. Similarly, very different notions of achievement and success introduced through colonisation fail to embrace a holistic style of learning whereby 'knowledge of indigenous heritage reinforces social attributes nurtured through relations with the land and environment, subsistence activity, social relationships, and responsibilities to other community members' (Wotherspoon, 2014: 326). Such a reorientation has had a major impact on the learning, development and cultural wellbeing of Indigenous children.

Linked to these over-arching changes from colonisation, a crucial loss for most Indigenous groups has been that of traditional philosophy and language. Philosophy and language have been pushed aside through colonisation to the extent that Western 'tools of analysis have meant that pre-colonial notions of reality have become submerged in the face of the English language' (Smith, 2000: 56). Amongst other things, for children this has resulted in many cultural practices, including games and songs being suppressed and lost to their development.

Equally, ongoing exploitation of Indigenous people's land, resources and intellectual property – dispossession as well as oppression – has been momentous in dissolving traditional institutions, practices and identities. Furthermore, directly related adverse environmental changes within their territories, due to different values of care, ownership and use, have compromised Indigenous peoples' long-held roles as guardians of the health of traditional lands and waters – places/spaces are often understood as intimately related ancestors, not simply physical forms. Lack of control over territories and resources impairs Indigenous groups' ability to develop in accordance with their own aspirations and to foster locational connectedness for the younger generations. This situation impinges upon ambitions to hand-on a secure identity that enhances the sense of belonging and pride that is elemental for the wellbeing of children of Indigenous heritage.

Despite some shifts away from colonial practices in most countries, there is a lingering legacy in terms of education and poverty in relation to Indigenous children, and in regard to Indigenous rights per se. The often partial attempts to bring about a 'post'-colonial society have been described by some commentators as a form of democratic colonialism – a process by which

> the contradictory ways in which formal adherence to democratic principles devoted to fostering equality in a pluralist society coexist with practices and discourses (many of them embedded or concealed within social and institutional structures) . . . perpetuate racism and exclusion of indigenous people.
> Wotherspoon, 2014: 328

So, for example,

> While formal guarantees are in place to honour indigenous rights and status and compensate for the damage of a colonial legacy, the realms in which these can be exercised in a meaningful way are constrained by practices driven by market priorities, policy factors and aspects of public discourse that reinforce differentials in power and privilege associated with that legacy.
> Wotherspoon, 2014: 336

This situation certainly means that fulfilling Indigenous visions is a constant uphill struggle. Nevertheless, Indigenous peoples have devised a variety of strategies to reclaim various degrees of self-determination in regard to coloniser-affected social structures, philosophical bases and traditional resources. The strategies are instructive both in their detail and broader story. This chapter cannot expand on the specifics of such undertakings, but it can summarise the primary principles behind moves to restore the cultural wellbeing of Indigenous children. The principles have potential for wider adaptation and application.

Principles for revitalising futures for Indigenous children

Any vision for the future necessarily acknowledges that 'all cultures (especially colonized ones) are perpetually in a state of change' (Kincheloe and Steinberg, 2008: 142). The key in this instance – while recognising such dynamism – is to ask what is paramount for achieving success in promoting wellbeing and resilience in Indigenous children? The representation in Figure 15.1 is a way of conceptualising the overarching messages delivered by scholars and practitioners working directly on issues surrounding the advancement of the cultural wellbeing of Indigenous children. This conceptualisation provides the foundation for discussion in this section.

The umbrella notion of Figure 15.1 is that it is crucial to create environments where Indigenous language, knowledge, culture and values are recognised as normal and legitimate; where being Indigenous is normal (Bishop *et al.*, 2003). The two principal and interweaving vertical components of Figure 15.1 in terms of the

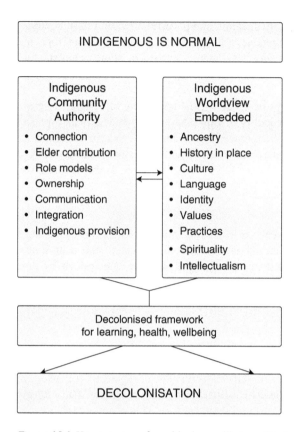

Figure 15.1 Key messages for achieving wellbeing of Indigenous children.

wellbeing of children are based on the pillars of Indigenous community control and honouring the Indigenous worldview of the people concerned, both of which are regarded as imperative to success in a cultural sense.

Starting with the left pillar, there is a significant body of commentators reporting that Indigenous community authority over service provision relevant to children is central to transforming health and education outcomes for Indigenous children (for example, Wotherspoon, 2014; MacDonald *et al.*, 2013; Sims *et al.*, 2008). This is recognition that the wellbeing of Indigenous children is interconnected with the wellbeing of their families and communities. Cultural identity develops where attachment is secure within their home community. A sense of community connectedness and responsibility for Indigenous children's wellbeing is therefore a starting place for achieving such aspirations.

Cultural wellbeing and belonging is nurtured when a child can play and develop within a familiar world. Of necessity, this will involve affirmation and involvement of family and community. Participation of Indigenous communities in education, for example, has been shown to increase student attendance, interest, participation and success (Whitinui, 2010). It also facilitates connection with positive role models and mentoring from elders in formal and informal settings. Community ownership of such activities and the associated community empowerment is decisive in fostering good outcomes. For instance, evidence suggests that community control via Indigenous leadership and ownership of education and health services for children results in consistently better service delivery and improved outcomes for children and their families (Besaw *et al.*, 2004; Secretariat of National Aboriginal and Islander Child Care, 2012). But it will be different in each community, with unique governance procedures, community values and practices, and management arrangements.

What is signalled here is the benefit of ongoing, in-depth communication that supports direct connection between the Indigenous community and their health and education service providers, and this is obviously more easily attained when the service is governed and managed by the Indigenous community. Trust in key people, from those involved in governance activities to those who actually deliver the service, is paramount. Those involved in any form of child care need to establish trust with the local Indigenous community, which dovetails with being in a culturally appropriate setting with Indigenous staff who are embedded in the local culture and language (Sims *et al.*, 2008). Furthermore, a holistic focus – which is commonly favoured by Indigenous groups – will mean prioritising the integration of services for children in areas such as education, health, cultural resources and housing (Sims *et al.*, 2008).

Turning now to the second pillar of Figure 15.1, it is clear that thinking based on each Indigenous community's own cultural referents is essential, enabling the conceptualising of development and wellbeing within their own worldview (Trask, 1993: 54). For example, for many Indigenous peoples the personification of the natural world is central to traditional culture and should be upheld in the raising of children. In the case of New Zealand Māori, for instance, relationship to natural features of the environment is not just legend, but personification of their ancestry, resulting from creation narratives and reinforced by the traditional practice of returning the placenta of each newborn to the earth of their homeland, thereby intimately connecting child, ancestors and place. Māori models of health, such as Te Whare Tapa Whā, enable a focus on identification, acknowledgement and incorporation of values and actions beyond the individual when considering the health and the everyday life of a child. Wellbeing can be powerfully attached to relationships with family, ancestors and the natural world through a deep-rooted experience of a particular place.

Cultural and language revitalisation that connects children to their learning and history in a way that is consistent with their worldview, and is not a Westernised abstraction of their traditions, is a key facet in such transformation. It is consistently reported that it is critical to nurture children as culturally connected learners: immersed in cultural identity and affiliation; conscious of cultural continuity; nestled in a shared heritage. Fitzsimons and Smith (2000), for example, stress the importance of passing on both interconnected notions of genealogy and current relationships for the formation of Indigenous identity and concomitantly

Indigenous wellbeing. Genuinely integrated and contextualised (in terms of landscapes, environments, events) use of the language, stories, music, art and symbols of their own culture is a necessary part of childhood education (Klopper, 2008). A case in point is reported in a review article related to youth suicide by MacDonald *et al.* (2013: 21775), which stressed that one of the key themes for Indigenous wellbeing was 'nurturing culturally specific relationships with the land, animals and plants, and developing cultural-based skills and activities'. Children are empowered by being strong in their culture and proud of their identity. This builds their confidence and self-esteem, positioning them positively in terms of their sense of belonging in community and broader engagement in society. Similarly, in an article on wellbeing among young Inuit in the Canadian Arctic, Kral *et al.* (2011: 433) found that some of the most important predictors of wellbeing related to 'traditional Inuit cultural values and practices'. It is important to cultivate 'indigenous ways of knowing and transmitting knowledge' (Sims *et al.*, 2008: 57); preferred ways of learning such as through elders, prayer, celebration and mentoring (Macfarlane *et al.*, 2007) can be quite different from Western convention. Space for distinctive Indigenous learning and intellectualism within education (that navigates beyond the confines of Western knowledge) that is shaped by Indigenous legacy and realities is central. There are a variety of routes by which to achieve this, including through the incorporation of traditional knowledge, values, practices, spirituality and activities in childhood learning and play. Ensuring learning and play is relevant to 'being Indigenous' is generally regarded as essential (Raerino, *et al.*, 2013). This means that more culturally aware practices of inclusion – based on Indigenous ways – are important in health and education decision making. Traditional games and songs, for instance, guide children through their social and spatial context and teach them varied and practical virtues and values – an interactive process of cultural apprenticeship (Nyota and Mapara, 2008). Children spend a great deal of time in school and such institutions therefore have a decisive role in fostering concrete learning relevant to Indigenous identity that allows children to cherish what is unique to their culture and place (Khasandi-Telewa *et al.*, 2012). In Aotearoa New Zealand, nurturing the performing art of Kapa Haka (dance in rows) has proved to be a route to fostering the wellbeing of Māori children, particularly through the intertwined link it has to engaging Indigenous children in 'learning more about their own language, culture and traditional ways of knowing and doing' (Whitinui, 2010: 24).

There are various recent examples of Indigenous groups proactively embracing such developmental priorities while working with national and local governments to achieve a diminishing of the wellbeing disparities that arose after colonisation. For instance, MacDonald *et al.* (2013: 21775) report favourable outcomes for Sami from:

> 30 years of explicit and conscious political, cultural and language revitalization efforts in Norway, which includes the development of many Sami institutions (i.e. schools, hospitals, a Parliament, and a University), a high degree of self-governance, good living conditions, increased soci-economic status and

positive socio-cultural development. These efforts and developments have contributed to supporting the Sami lifestyle and culture allowing Sami youth more opportunities to learn and participate in their culture, as well as to assist in improving health outcomes.

Nevertheless, it is more than reviving Indigenous ways that is necessary; it is defending and decolonising them and developing new frameworks for achieving a move forward (Waerea-I-te-rangi Smith, 2000). For example, in Aotearoa New Zealand, the 1980s development of immersion language pre-schools and primary schools (Te Kohanga Reo and Kura Kaupapa Māori) as an alternative to conventional schooling was a concerted effort to achieve such revitalisation, with considerable success – and with a fundamental contribution to the survival of Māori culture and structures long term (Smith, 1997). Contemporaneously, any restructuring of the educational and wellbeing circumstances for Indigenous children needs to be mirrored by comprehensive decolonisation strategies integrally linked to recognition of Indigenous sovereignty. A significant change in outcomes will not be achieved by merely tinkering with the current flawed systems (Sims *et al.*, 2008).

Conclusion

Indigenous populations in (post)-colonial settings have very different contexts and needs to the dominant population. Mainstream policies and strategies rarely meet such needs effectively. Although emphasising the importance of not homogenising Indigenous values and aspirations, this chapter points towards an ambition of embracing Indigenous cultural, economic, social, spiritual and environmental knowledges and resources in order to shift from a privileging of the majority to a co-existence of self-determining peoples. Any such shift in favour of genuine equity of participation in society by Indigenous groups has a concomitant need for the nurturing of leadership and the empowerment of Indigenous autonomy. Research on Indigenous initiatives demonstrates there is much to be gained by empowering Indigenous communities to facilitate and deliver wellbeing, education and play that is contextualised to place and identity/ies, in accordance with their own cultural aspirations – geographically positioned, historically embedded, holistically interconnected, consciously specific, but also continually negotiated. Such Indigenous transformation also has implications for dominant society, which needs to be informed, even reformed, by the challenges emerging from the Indigenous world.

References

Besaw, A., Kalt, J., Lee, A., Sethi, J., Boatright Wilson, J. and Zemler, J. 2004. *The context and meaning of family strengthening in Indian America*. A Report to the Annie E. Casey Foundation by The Harvard Project on American Indian Economic Development.

Bishop, R., Berryman, M., Tiakiwai, S. and Richardson, C. 2003. *Te Kōhtahitanga: the experience of Year 9 and 10 Māori students in mainstream classrooms*. Rangahau Mātauranga Māori, Māori Education Research Report. Wellington: Ministry of Education.

Borell, B., Gregory, A., McCreanor, T., Jensen, V. and Moewaka Barnes, H. 2009. 'It's hard at the top but it's a whole lot easier than being at the bottom': the role of privilege in understanding disparities in Aotearoa/New Zealand. *Race/Ethnicity*, 3(1), pp. 29–50.

Clark, S. 2009. Editorial: where are we now with Indigenous health? *Lancet*, 374, p. 2.

Durie, M. 2004. An Indigenous model of health promotion. *Health Promotion Journal of Australia*, 15, pp. 181–5.

Fitzsimons, P. and Smith, G. 2000. Philosophy and Indigenous cultural transformation. *Educational Philosophy and Theory*, 2(1), pp. 25–41.

Gracey, M. and King, M. 2009. Indigenous health part 1: determinants and disease patterns. *Lancet*, 374, pp. 65–75.

Herman, R.D. 2008. Reflections on the importance of Indigenous geography. *American Indian Culture and Research Journal*, 32, pp. 73-88.

Khasandi-Telewa, V., Liguyani, R. and Wandera-Simwa, S. 2012. Appropriating globalisation to revitalise indigenous knowledge and identity through Luhya children's play songs. *The Journal of Pan African Studies*, 5(6), pp. 75–91.

Kincheloe, J. and Steinberg, S. 2008. Indigenous knowledges in education: complexities, dangers, and profound benefits. In Denzin, N., Lincoln, Y. and Smith, L. (eds.) *Handbook of critical and indigenous methodologies*. Thousand Oaks, CA: SAGE Publications, pp. 135–57.

King, M., Smith, A. and Gracey, M. 2009. Indigenous health part 2: the underlying causes of the health gap. *Lancet*, 374, pp. 76–85.

Klopper, C. 2008. Meeting the goals of Te Whariki through music in the early childhood curriculum. *Australian Journal of Early Childhood*, 33(1), pp. 1–8.

Kral, M., Idlout, L., Minore, J., Dyck, R. and Kirmayer, L. 2011. Unikkaartuit: meanings of well-being, unhappiness, health, and community change among Inuit in Nunavut, Canada. *American Journal of Community Psychology*, 48, pp. 426–38.

MacDonald, J., Ford, J., Willox, A. and Ross, N. 2013. A review of protective factors and causal mechanisms that enhance the mental health of Indigenous Circumpolar youth. *International Journal of Circumpolar Health*, 72, pp. 21775.

Macfarlane, A., Glynn, T., Cavanagh, T. and Bateman, S. 2007. Creating culturally-safe schools for Maori students. *The Australian Journal of Indigenous Education* 36, pp. 65–76.

Martin, B. 2000. Place: an ethics of cultural difference and location. *Educational Philosophy and Theory*, 31(1), pp. 81–91.

Nyota, S. and Mapara, J. 2008. Shona traditional children's games and play: songs as indigenous ways of knowing. *The Journal of Pan African Studies*, 2(4), pp. 189–202.

Peters, M. 2000. Editorial. *Educational Philosophy and Theory*, 32(1), pp. 5–13.

Raerino, K., Macmillan, A. and Jones, R. 2013. Indigenous Māori perspectives on urban transport patterns linked to health and wellbeing. *Health and Place*, 23, pp. 54–62.

Robson, B. and Harris, R. (eds.). 2007. *Hauora: Māori standards of health IV. A study of the years 2000-2005*. Wellington: Te Rōpū Rangahau Hauora a Eru Pōmare.

Secretariat of National Aboriginal and Islander Child Care (SNAICC). 2012. *Improved outcomes for Aboriginal and Torres Strait Islander children and families in early childhood education and care services: learning from good practice*. Melbourne: SNAICC.

Sims, M., Saggers, S., Hutchins, T., Guilfoyle, A., Targowska, A. and Jackiewicz, S. 2008. Indigenous child care – leading the way. *Australian Journal of Early Childhood*, 33(1), pp. 56–60.

Skinner, E. and Masuda, J.R. 2013. Right to a healthy city? Examining the relationship between urban space and health inequity by Aboriginal youth artist-activists in Winnipeg. *Social Science & Medicine*, 91, pp. 210–18.

Smith, G. 1997. The development of Kaupapa Maori: theory and praxis. PhD thesis, University of Auckland, New Zealand.

Smith, T. 2000. Nga Tini Ahuatanga o Whakapapa Korero. *Education Philosophy and Theory*, 32(1), pp. 53–60.

Te Aho, L. 2011. Indigenous aspirations and ecological integrity: restoring and protecting the health and wellbeing of an ancestral river for future generations in Aotearoa New Zealand. In: Westra, L., Boseelmann, K. and Soskolne, C. (eds.), *Globalisation and ecological integrity in science and law*. Newcastle-Upon-Tyne: Cambridge Scholars, pp. 346–60.

Tobias, M., Blakely, T., Matheson, D., Rasanathan, K. and Atkinson, J. 2009. Changing trends in indigenous inequalities in mortality: lessons from New Zealand. *International Journal of Epidemiology*, 38, pp. 1711–22.

Trask, H. 1993. *From a Native daughter: colonialism and sovereignty in Hawaii*. Monroe, ME: Common Courage Press.

Waerea-I-te-rangi Smith, C. 2000. Straying beyond the boundaries of belief: Maori epistemologies inside the curriculum. *Educational Philosophy and Theory*, 32(1), pp. 43–51.

Whitinui, P. 2010. Kapa Haka 'voices': exploring the educational benefits of a culturally responsive learning environment in four New Zealand mainstream secondary schools. *Learning Communities: International Journal of Learning in Social Contexts*, 1, pp. 24–54.

Wotherspoon, T. 2014. Seeking reform of Indigenous education in Canada: democratic progress or democratic colonialism. *Alternative*, 10(4), pp. 323–39.

Young, I.M. 1995. Together in difference: transforming the logic of group political conflict. In Kymlicka, W. (ed.), *The rights of minority cultures*. Oxford: Oxford University Press, pp. 155–78.

16 Ecological wellbeing, childhood and environmental change

Ann E. Bartos and Bronwyn E. Wood

For decades, children have been deployed as symbols of the environmental movement. Whether as speakers at important global meetings, as photogenic participants at protests holding handmade signs, or as the ill-fated 'future generations' whose destiny lies in the hands of adults' poor decision making; children and childhood are often strategically and politically utilised to bring attention to environmental problems. Through these innocuous forms of manipulation, children are situated in a paradoxical position as both future saviours and the most vulnerable victims of the impending dystopic future they are 'doomed' to inherit.

However, both these positions of victim and saviour fail to recognise the complexity of children's wellbeing. Children are positioned conversely as passive or vulnerable in the face of rapid environmental or societal change, or as autonomous agents able to 'fix' these complex issues. In this chapter we advance an alternative position that focuses on the notion of 'ecological wellbeing' in an attempt to recognise the interconnectedness of social, political, economic and environmental factors on the wellbeing of children and, in turn, their role in shaping and responding to such factors. Building on a small number of studies that have taken such an approach, our aim is to more fully explore the complex micro and macro processes that play an important role in children's ecological worlds (Katz, 1994; Chawla, 2007; Hayward, 2012; Carroll *et al.*, 2015). This is not a simple task. As Hayward (2012) argues, the 'environment' and 'ecology' are fluid concepts for children which makes it challenging to both investigate and understand. It is important to do so, however, because 'we rarely take the complex ecological reality of a child's world's seriously [. . . as] their ecological truths are a little too inconvenient' (Hayward, 2012: 3). In an effort to take children's worlds seriously, in this chapter we aim to tease out a number of ways in which children's 'ecological truths' and notions of wellbeing intersect.

To that effect, we introduce the concept of 'ecological wellbeing' as a conceptual tool to assist in untangling the social, political, economic and environmental interrelationships that work in tandem to promote or jeopardise a sense of childhood wellbeing. We build our understanding of wellbeing in line with the shift from biomedical disease models to broader multi-disciplinary social models of health and health care (Kearns and Andrews, 2010; Morrow and Mayall, 2010). Kearns and Andrews (2010) introduce a more expansive notion of wellbeing to seek a deeper understanding of how the emotional terrain and the cultural importance of place intersect in cultural, social and political ways with significant implications for physical and emotional landscapes of health care. Such developments have led to more comprehensive notions and understandings of both wellbeing and the environment itself, which inform our research. Furthermore, we place this expansive and non-medicalised notion of wellbeing within holistic, 'nested' and ecological understandings of society; ecological approaches highlight the interconnectedness and interconnectivity of humans and the natural world (Dobson, 2003; Hayward, 2012; Metcalfe and Game, 2014). Situating this approach against a backdrop of environmental change and degradation provides the context by which we interrogate the concept of 'ecological wellbeing' more closely.

Our central argument in this paper is that our understanding of children's wellbeing is greatly enhanced when we place it within an ecological framework. Although this combination of concepts (ecology + wellbeing) has received scant attention in geography to date, we believe that the conceptualisation of an ecological wellbeing framework provides a valuable tool for researchers of children and youth to consider the import of place, power and interrelationality in shaping children's wellbeing, along with their interaction and engagement with environments. We outline some of the more relevant discourses of both wellbeing and ecology in this chapter to promote a provocative coupling of the two concepts in an effort to better understand the active role children have in creating the environment they want now and in the future.

We begin with an examination of literature interrogating issues of environmental change and how this affects children's wellbeing. We then examine how ecological metaphors have been employed in various childhood studies, which highlight the import of the socially and politically embedded nature of children's interactions, experiences and responses to their environments. Together, this literature contributes to an integrated framework to advance the concept of *ecological wellbeing*, for which we outline three interrelated premises: first, that an ecological wellbeing approach highlights an *interconnected* understanding of children as members of families, communities and institutions as well as environments; second, that such an approach recognises the significance of *competing power relationships* within society and how these play out upon children's lives; and third, that children's wellbeing is not something they passively inherit, but that they themselves contribute to their own wellbeing and that of others. We conclude with a discussion about how an ecological approach to wellbeing can enhance understandings of wellbeing in general and for children in particular.

Environmental change and children

Scholars of global environmental change have drawn attention to how human populations increasingly negotiate a new politics of uncertainty related to migration (Biermann and Boas, 2010), national border security (Barnett and Adger, 2007; Adger, 2010), structural inequality and land use (Boston, 2009), and health concerns (Costello *et al.*, 2009). These uncertainties impact social and environmental relationships and livelihoods, and ultimately wellbeing (Curtis and Oven, 2012). Curtis and Oven (2012) argue that children are among the 'subpopulations with greater vulnerability' to the health effects of climate change, particularly those who are also socio-economically disadvantaged. As a particularly 'vulnerable' population in the face of impending environmental degradation, researchers argue that children in the majority world will face an increasing number of social and physical stresses. Physical stresses may be caused by limited access to scarce resources, susceptibility to communicable diseases, impacts of extreme weather, and increased cases of malnutrition. Psychological stresses may result from migration, loss of economic or political security, decreased access to education, and significant changes in social reproduction (UNICEF, 2011).

Within minority world urban environments, children frequently experience environmental change as a result of changing patterns in the built environment. For example, growing levels of urbanisation and intensive urban design often lead to a reduction in access to outdoor everyday environments (Chawla, 2001; Kearns *et al.*, 2003; Freeman and Tranter, 2011; Freeman, 2013). Research has demonstrated that as families become more car-dependent, children suffer a decrease in independent mobility and have less time for outdoor play (Kearns *et al.*, 2003). However, although there is strong evidence to suggest that these changes have been primarily negative, Freeman and Tranter (2011) suggest that it is important to also see how children experience new opportunities within these constraints. For example, Freeman (2013) highlights the opportunities for wider spatiality of participation in cities that modern forms of car travel and public transport offer children, as well as the ongoing commitment of many schools and families to expose children to the outdoors and natural environments. Such research suggests that we need to recognise how children themselves are responding to environmental change. Insights such as those found by Freeman (2013) have the potential to uncover alternative avenues for expressions of children's agency that challenge the assumption that children are victims or a vulnerable population unable to cope with the changes underway through global environmental change. This approach acknowledges children's everyday engagements in their local environments and how this everyday engagement enables them to cope with and adapt to environmental change beyond the structural accounts of access to clean drinking water, hygienic sanitation or medical facilities, for example.

The research cited earlier highlights the need to adopt a more holistic framework of 'wellbeing' to include consideration of how personal and structural dimensions impact everyday matters of wellbeing and attention to children's agency in negotiating environmental change on a daily basis. Additionally, we believe that broadening the focus on wellbeing to include the embodied, emotional and relational dimensions of how children relate to their environments helps to bring attention to their active role in the creation of the worlds they want to live in now and in the future. We discuss this point later in the chapter. The following section explores our understanding of ecological metaphors to uncover a more nuanced approach to studies of wellbeing.

Ecological metaphors in childhood research

Ecology is a branch of biology that investigates the relationship between organisms and their surrounding environments. Social scientists use the metaphor of ecology to bring attention to the interconnections that exist between seemingly disparate things, people or places due to a variety of social, political and environmental phenomenon. Within education and childhood studies, Bronfenbrenner's (1974, 1979, 1994) ecological model of human development is frequently drawn upon to explain research on children's wellbeing (Amerijckx and Humblet, 2014). Bronfenbrenner suggests that human behaviour can only be understood by examining 'multiperson systems of interaction not limited to a single setting and must take into account aspects of the environment beyond the immediate situation containing the subject' (1974: 514). He describes the ecological environment as a nested arrangement of concentric structures, each contained within the next – and referred to as micro, meso, exo and macrosystems (Bronfenbrenner, 1979: 22). He suggests that human development needs to be examined and evaluated within each of these structures, beginning with the micro (interpersonal relations) and extending to mesosystem settings (such as school, the home, workplace), exosystems (the influence of government agencies, transport systems) and macrosystem settings (societal ideologies) (see Figure 16.1).

Although Bronfenbrenner's model provides a useful framework for understanding the complexity of relationships within the social context of everyday life, it overlooks the political and power-laden nature of these relationships (France *et al.*, 2012). France and colleagues propose a 'political ecology' approach to Bronfenbrenner's original ecological model by more explicitly embedding Bourdieu's concepts of power (c.f. Robbins, 2004). In their work, they recognise 'that the everyday "worlds" that young people engage in, and interact with, are a product of external 'political' forces evident at a number of levels (within microsystems, mesosystems, exosystems and macrosystems)' (France *et al.*, 2012: 5). For example, they demonstrate how young people's criminal identities and encounters with criminal activities in their study were not only a product of developmental and psychological factors at an individual level, but were strongly influenced by experiences of poverty, social exclusion and social inequalities operating at a wider level. Rather than viewing children and their actions in a state of isolation or disconnection, or assuming that they hold equal or homogenous opportunities for participation, France *et al.* (2012) advance upon Bronfenbrenner's model by highlighting the significance of embedded power relations within children's experiences and interactions. Applying this 'political ecology' approach to investigations of wellbeing provides an opportunity to explore the complex interactions between individuals and their social, cultural and economic contexts in a way that exposes the imbalances of power operating across ecological systems.

Ecological approaches have also been useful to some children and youth scholars within geography, most notably Jeffrey (2008, 2010, 2013) and Katz (1994, 1998, 2004). In their research focused on youth in India, and Africa and Harlem, respectively, they investigated the ways that global economic conditions changed and

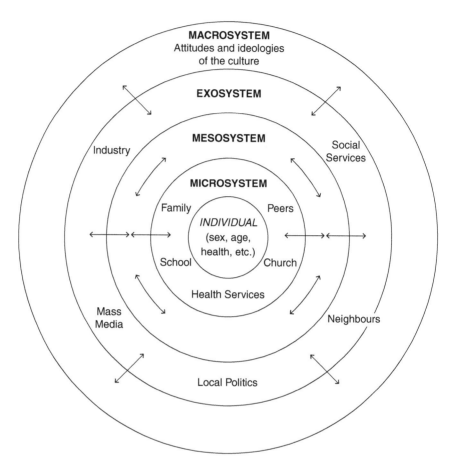

Figure 16.1 Bronfenbrenner's ecological systems theory shown in a diagram.

Source: cited on http://en.wikipedia.org/wiki/Ecological_systems_theory#The_five_systems (accessed 22 August 2014).

abruptly altered young people's everyday survival and transition into adulthood while also contributing to 'novel cultural, political, and social strategies' among youth (Jeffrey, 2008: 753). These new strategies are a result of what Jeffrey identifies as 'new ecologies of youth' composed of significant changes to the technological, built and natural environment of today's youth around the world. He argues that,

> [t]he ecological metaphor does triple service. It highlights how technologies, materials and specific places are mutually entangled in the production of oppositional politics. It hints, too, at the complexity, spontaneity and mutability of action. The ecological metaphor [. . .] can provide a framework for examining the rhythms, regulations and improvisations that constitute politics in practice.
>
> Jeffrey, 2013: 150

Similarly, Katz draws on political economic theory to explore what she has called the 'political ecology of childhood'. For Katz 'the political ecology of childhood is both a constellation of ideas and a set of material circumstances that frame children and the environment as a pivotal social, political, economic and cultural relation' (1994: 108). She understands the environment to be both socially constructed and produced as well as a physical space that both promotes and inhibits children's wellbeing. According to Katz, a political ecology of childhood would

> call for an environment that provides for children's health and wellbeing; safety; physical, emotional and intellectual development; and future, broadly understood ... [C]*hildren's health and wellbeing* depend, for example, on adequate food, shelter, health care, sanitation, social services and protection from environmental pollution and hazards.
>
> Katz, 1994: 108, emphasis added

Drawing on the variety of perspectives of (political) ecology adopted by such scholars as those cited earlier enables us to suggest three premises that underlie our concept of ecological wellbeing. Our first premise begins with focused attention to the interconnected and nested nature of children's everyday experiences that influence notions of wellbeing. This ecological approach to wellbeing allows us to interrogate how the material and immaterial aspects of the environment influence children's lives in ways that have emotional, relational and spatial implications for their wellbeing. Our second premise recognises that these experiences and relationships are not neutral, but reflect wider social and political power relationships within society. Incidentally, as children respond to both environmental and human-related changes within their worlds, we recognise that notions of wellbeing also change. Furthermore, rather than viewing wellbeing as something that is passively received in relation to the environment, our third premise highlights how children actively create and contribute to aspects of their own and others' wellbeing. In the following section, we examine each of the three premises underpinning our understanding of ecological wellbeing in greater detail, providing illustrations from our own and others' work.

An ecological wellbeing approach to childhood: three premises

Our first premise is that an ecological approach highlights a strongly interconnected understanding of children's wellbeing, positioned within multiple interrelationships between children and groups around them (family, friends, communities and institutional relationships) and environments. Children's wellbeing cannot be isolated from these relationships, nor from the material environmental factors within their neighbourhoods and communities (Hart, 1979; den Besten, 2010; Ben-Arieh *et al.*, 2014; McKendrick, 2014). Instead, wellbeing needs to be understood as a multi-dimensional and multi-faceted concept (Ben-Arieh *et al.*, 2014) that is deeply embedded within micro and macro geographies that extend well beyond narrow psychological notions of wellbeing and are more aligned with the expansive approach to wellbeing outlined by Kearns and Andrews (2010).

In our own work in New Zealand we have clearly seen this multi-dimensional nature of wellbeing (Bartos, 2011, 2013; Wood, 2012, 2013). In our separate research projects, we explored young people's perceptions of 'belonging' to a neighbourhood in the context of young citizens (Wood) and descriptions of 'special places' in their everyday lives (Bartos). In our research projects, we found that the children were developing a sense of wellbeing that was often deeply tied to their relationship with their physical and social environments. Although these were often initially perceived on a microscale (using Bronfenbrenner's terminology), their responses also reflected real or potential threats to these environments on a wider macroscale. For example, in Bartos's (2011) and Wood's (2013) studies, many young people identified 'special' sites representing a sense of belonging, which were notable for their familiarity and everydayness, such as trees, local streets, local shops, playgrounds and homes. The inclusion of friends and family members were nearly always evident in their descriptions of the 'specialness' of the sites. When asked why these were important, replies typically emphasised spatial practice and temporality: 'I've been going [to that shop] since I was little', or 'My granddad planted that tree' and 'I've lived there my whole life'. Although the physical site may have had some inherent value to the children we worked with, much of what they articulated to be special in these mundane sites had much to do with a sense of attachment infused with contexts of social interaction and memory (de Certeau, 1984). Their interrelationships with their friends and family significantly impacted their sense of belonging at the microscale.

Such descriptions painted a picture of ecological wellbeing that was centred on everyday interactions and familiar spatial and social practices, which were strongly affective and relational. However, their sense of belonging also extends further towards the macroscales when young people saw their 'special' places under threat by urban encroachment, pollution, graffiti or environmental degradation. Their desire to care, protect or nurture such places reinforced how these places contributed to their own wellbeing and the wellbeing of others. An ecological wellbeing approach acknowledges that children's wellbeing is reflective of the interconnections evident within the micro to macroscales. Our examples highlight these interconnections through the dimension of belonging, which we believe contributes to a sense of wellbeing; other dimensions of wellbeing such as health, safety, pleasure and freedom should also be explored at various scales to shed light on the importance of interactions from the personal to the structural (for an adult context, see Fleuret and Atkinson, 2007; Atkinson *et al.*, 2012; Chapter 13, this volume).

Our second premise is that an ecological wellbeing approach recognises that the relationship between children's wellbeing and the environment is not neutral, but reflects wider social and political power relationships within society. Children are often caught between competing agendas within these relationships. Katz's (2004) research on the changing political ecologies of Howa, Sudan provides an excellent example to explore this premise. The physical nature of Howa underwent significant ecological changes when the state-sponsored and internationally financed Suki Agricultural Project was introduced into the predominantly subsistence-based economy in the 1980s. Through moving to a cash-crop economy, the village suffered from increased ecological destruction necessary for the new agricultural economy to be productive and also from severe changes to traditional forms of social reproduction, gender roles and childhood. Children were frequently required to participate in the new economy through their labour, which took them away from their schooling and also left them unprepared for a future in their village, which now requires land ownership and inheritance. These complex relationships between children's experiences and knowledge production of their environment and the underlying social, political and economic structures that deem certain experiences and knowledges to be moot had a significant impact on children's wellbeing in Katz's research site.

Hayward (2012) suggests that such examples are becoming much more widespread in the majority world under neoliberal economic restructuring projects. Children and young people are part of a global cohort who are exposed to increasing levels of child poverty, environmental risk and intergenerational inequity, which has a profound effect on their wellbeing. She argues that at such times, a political dimension and a commitment to democracy is increasingly relevant to counter the deeply ecological nature of such problems. We are beginning to witness such commitment of young people across the Global North and Global South through the Occupy Movements, the Arab Spring, the anti-globalisation movements and the climate justice movements in which young people raise concerns that tie local environmental concerns with larger structural issues in an effort to increase equality and sustainability.

The children in Bartos's (2011) research in the minority world were similarly influenced by the political economy of New Zealand's dairy industry and the role of the dairy economy in their daily lives. At school, the roll changed dramatically during the milking season when parents, with their children, migrate looking for work. This required, for example, the children to be resilient and adept at making new friends throughout the year, whether they are the ones staying or moving on. At home, children from dairying families had a variety of roles and responsibilities after school including farm chores and babysitting younger siblings while their parents worked long hours on the farm. These children developed a sense of responsibility in the home which was often transferred to the classroom and to their friendship networks. Within the larger community, the children would participate in community-driven ecological riparian restoration on local farms. Such activities required the children to physically participate in restoring the health of the streams that are degraded from intensive commercial dairying. Through bringing attention to their local geographies, it was evident that their personal lives were intricately linked with the global economy (see also Kearns *et al.*, 2009). Whether in the school, the home, or the larger community, the dairy industry played an important role in many of the children's lives and influenced how the children actively participated in building friendships, establishing responsibility and learning about (un)sustainable farming practices. Their sense of wellbeing is ecological in that it is largely influenced by the wider political economy of the dairy industry. Such examples (Katz, 2004; Bartos, 2011; Hayward, 2012) support our premise that an understanding of ecological wellbeing is both complicated and elucidated through focused attention towards the micro to macroscales of analysis and their intersections (e.g. Bronfenbrenner).

Our third premise underpinning ecological wellbeing views wellbeing as something that is not passively received or inherited by children, but something that they actively contribute to, thus enhancing their own wellbeing and that of others. One way to uncover children's active agency in the creation of wellbeing is through a focus on children's embodied, emotional and sensorial everyday experiences and practices. These more mundane and 'unspectacular' forms of agency can shed light on how children negotiate complex ecological and social conditions that may be otherwise unacknowledged in the research; children's descriptions of embodied sensations of being (un)comfortably cold or hot, wet or dry, and content or scared in physical spaces can shed light on how children's sense of wellbeing is often very personal and intimately felt and sensed (see also Ergler *et al.*, 2013). For example, the children in Bartos's (2013) ethnographic research described their positive and negative responses to their physical environment based on how their bodies felt in these places. One child described how his favourite spot to sit on his property allowed him time to look at his environment and appreciate its subtleties such as the weather, the vegetation, the biodiversity and the social interactions going on around him. Through the ordinary and mundane act of sitting, he was able to use a variety of his senses (the microscale) to engage with his environment (the macroscale). This engagement can contribute to a more complex understanding of how ecological wellbeing is embodied in everyday practices and has the potential to encourage a sense of appreciation for the physical environment based on feelings and engagements on a visceral level. Appreciation for their physical environment has the potential to garner a sense of responsibility to maintain and establish the type of living situations that in turn promote and nurture children's wellbeing.

Wood's (2012) research with a group of Pasifika young people in a low socio-economic suburb also illustrated the active role children and young people play in contributing towards their own wellbeing and that of others from the micro to the macroscales. These young people described how their wellbeing was derived from their experience of belonging in their multicultural community where diverse cultural groups mixed together well and social groups were respected and valued. They stated that this led to a supportive 'spirit' in the community. However, this strong sense of belonging was marred by a disappointment in the prevalence of alcohol outlets in their suburb and the aggressive marketing of cheap alcohol. Their concerns centred particularly on children and young people and on how cheap and easy access to alcohol contributed to truancy, unemployment and a number of social ills. Their insights were closely attuned to their own experience of being young and their concerns about their friends and family members.

They also recounted with pride how a local protest of families (including a number of their friends) had successfully prevented one further liquor outlet from being established in their suburb. The young people's indignation at the presence of such liquor outlets in their community showed their growing geographies of responsibility (Massey, 2004) as young citizens. This stemmed from a relational and emotional sense of wellbeing in their neighbourhood and a desire to protect and preserve what was special about this community.

Examples of ecological wellbeing at the macroscale are evident when exploring children's political geographies; political actions of children and young people that contribute to individual and societal wellbeing tend not to be restricted to individual or community issues. Hayward's (2012) examples of New Zealand children working toward the rebuilding of Christchurch with their families and communities following the Canterbury earthquakes illustrate how their geographies of responsibility stretched in scope and scale beyond notions of personal wellbeing to connect with a much deeper and ecological sense of wellbeing and place-making at a city and regional level. We believe Hayward (2012) and our own research helps demonstrate how children are actively involved in the creation of their wellbeing through their embodied and emotional reactions to ordinary, mundane and otherwise potentially ignored everyday situations.

Conclusion

In this chapter we have traced some of the concepts and ideas that contribute towards what we have termed an 'ecological wellbeing' of childhood. Set against the backdrop of environmental change and degradation, we have argued that an ecological wellbeing approach provides us with an expansive notion of wellbeing that draws attention to the interconnectedness of children's wellbeing to place, power and interrelationality. An ecological wellbeing approach also draws attention to both micro and macroscale aspects of wellbeing, thus connecting material, behavioural and affective characteristics with structural concerns (Kearns and Andrews, 2010; see Bronfenbrenner, 1974, 1979, 1994). We developed this framework through bringing attention to relationality, inherent power dynamics, and children's deliberate agency to demonstrate that a more ecological and encompassing understanding of wellbeing is necessary within childhood. Furthermore, examining the affective nature of emotional attachment and connection to places provides a sense of the 'emotional, cultural and imaginative ties to place' (Ansell and van Blerk, 2007: 18) that powerfully shape how children form their social relationships and social identities, thus giving further insights into children's ecological wellbeing.

Our understanding of an ecological wellbeing approach recognises that the everyday worlds of young people are products of external political forces evident at a variety of scales (i.e. Bronfenbrenner's concepts of microsystems, mesosystems, exosystems and macrosystems). Moreover, our approach pays particular attention to how children and young people themselves contribute to their own ecological wellbeing within the constraints imposed by adult society. Attention to how such place-attachments occur within specific spaces highlights the dynamic nature of both environments and ecological wellbeing, and how these are subject to change through time because they are never static.

In sum, we believe that the conceptualisation of an 'ecological wellbeing' framework provides a valuable tool for geographers and childhood scholars to consider the import of place, power and interrelationality in shaping children's wellbeing. We have proposed that an ecological wellbeing approach positions children within much wider and interrelated structures and processes in their environments and recognises the interconnected dimensions of a child's state of wellbeing. Children's ecological wellbeing can therefore be seen as a measure of the relative harmony or dissonance that children experience in their social and physical environments. This is particularly relevant in the contexts of both environmental and social change which can and do have significant impacts on children's everyday lives and their sense of wellbeing. Viewing wellbeing as a state of balance within nested and interrelated ecologies shifts the focus away from children as autonomous players and situates them within interconnected relationships within social and political networks and environments, thus providing more expansive notions and understandings of both wellbeing and ecologies.

References

Adger, W. N. 2010. Climate change, human well-being and insecurity. *New Political Economy*, 15(2), 275–292.

Amerijckx, G. and Humblet, P. C. 2014. Child well-being: what does it mean? *Children and Society*, 28(5), 404–15.

Ansell, N. and van Blerk, L. 2007. Doing and belonging: towards a more-than-representational account of young migrant identities in Lesotho and Malawi. In R. Panelli, S. Punch and E. Robsen, eds. *Global perspectives on rural childhood and youth: young rural lives*. New York: Routledge, pp. 17–28.

Atkinson, S., Fuller, S. and Painter, J. 2012. *Wellbeing and place*. Farnham: Ashgate.

Barnett, J. and Adger, W. N. 2007. Climate change, human security and violent conflict. *Political Geography*, 26, pp. 639–55.

Bartos, A. E. 2011. Remembering, sensing and caring for their worlds: children's environmental politics in a rural New Zealand town. PhD thesis. University of Washington, Seattle.

Bartos, A. E. 2013. Children sensing place. *Emotion, Space and Society*, 9(1), 89–98.

Ben-Arieh, A., Casas, F., Frones, I. and Korbin, J. 2014. *Handbook of child well-being: theories, methods and policies in global perspective*. Dordrecht: Springer.

Biermann, F. and Boas, I. 2010. Preparing for a warmer world: towards a global governance system to protect climate refugees. *Global Environmental Politics*, 10(1), 60–88.

Boston, J. 2009. *Eliminating world poverty: global goals and regional progress*. Wellington: Institute of Policy Studies.

Bronfenbrenner, U. 1974. Developmental research, public policy and the ecology of children. *Child Development*, 45(1), 1–5.

Bronfenbrenner, U. 1979. *The ecology of human development: experiments by nature and design*. Cambridge, MA: Harvard University Press.

Bronfenbrenner, U. 1994. Ecological models of human development. In T. Husen and T. N. Postlethwaite, eds., *International encyclopaedia of education* (Vol. 3, 2nd edn.) Oxford: Elsevier Sciences, pp. 1643–47.

Carroll, P., Witten, K., Kearns, R. and Donovan, P. 2015. Kids in the city: children's use and experiences of urban neighbourhoods in Auckland, New Zealand. *Journal of Urban Design*, 20(4), 417–36.

Chawla, L. 2001. Cities for human development. In L. Chawla, ed., *Growing up in an urbanising world*. London: Earthscan/UNESCO, pp. 15–34.

Chawla, L. 2007. Childhood experiences associated with care for the natural world: a theoretical framework for empirical results. *Children, Youth and Environments*, 17(4), pp. 144–70.

Costello, A., Abbas, M., Allen, A., Ball, S., Bell, S., Bellamy, R., Friel, S., Groce, N., Johnson, A., Kett, M., Lee, M., Levy, C., Maslin, M., McCoy, D., McGuire, B., Montgomery, H., Napier, D., Pagel, C., Patel, J., Puppim de Oliveira, J. A., Redclift, N., Rees, H., Rogger, D., Scott, J., Stephenson, J., Twigg, J., Wolff, J. and Patterson, C. 2009. Managing the health effects of climate change. *Lancet*, 373(9676), pp. 1693–733.

Curtis, S. E. and Oven, K. J. 2012. Geographies of health and climate change. *Progress in Human Geography*, 36(5), pp. 654–66.

de Certeau, M. 1984. *The practice of everyday life* (translated by S. Rendall). Berkeley, CA: University of California Press.

den Besten, O. 2010. Local belonging and 'geographies of emotions': Immigrant children's experience of their neighbourhoods in Paris and Berlin. *Childhood*, 17(2), pp. 181–95.

Dobson, A. 2003. *Citizenship and the environment*. Oxford: Oxford University Press.

Ergler, C. R., Kearns, R. A. and Witten, K. 2013. Seasonal and locational variations in children's play: implications for wellbeing. *Social Science & Medicine*, 91, pp. 178–85.

Fleuret, S. and Atkinson, S. 2007. Wellbeing, health and geography: a critical review and research agenda. *New Zealand Geographer*, 63, pp. 106–18.

France, A., Bottrell, D. and Armstrong, D. 2012. *A political ecology of youth and crime*. Basingstoke: Palgrave.

Freeman, C. 2013. The changing environmental worlds of Aotearoa New Zealand children. In N. Higgins and C. Freeman, eds., *Childhoods: growing up in Aotearoa New Zealand*. Wellington: Otago University Press, pp. 59–76.

Freeman, C. and Tranter, P. 2011. *Children and their urban environment: changing worlds*. London: Earthscan.

Hart, R. 1979. *Children's experience of place*. New York: Irvington.

Hayward, B. 2012. *Children, citizenship and environment: nurturing a democratic imagination in a changing world*. New York: Routledge.

Jeffrey, C. 2008. 'Generation nowhere': rethinking youth through the lens of unemployed young men. *Progress in Human Geography*, 32(6), pp. 739–758.

Jeffrey, C. 2010. Geographies of children and youth I: eroding maps of life. *Progress in Human Geography*, 34(4), pp. 496–505.

Jeffrey, C. 2013. Geographies of children and youth III: alchemists of the revolution? *Progress in Human Geography*, 37(1), pp. 145–52.

Katz, C. 1994. Textures of global change: eroding ecologies of childhood in New York and Sudan. *Childhood*, 2(1–2), pp. 103–10.

Katz, C. 1998. Disintegrating developments: global economic restructuring and the eroding of ecologies of youth. In T. Skelton and G. Valentine, eds., *Cool places: geographies of youth cultures*. London: Routledge, pp. 130–44.

Katz, C. 2004. *Growing up global: economic restructuring and children's everyday lives.* Minneapolis, MN: University of Minnesota Press.

Kearns, R. A. and Andrews, G. J. 2010. Geographies of wellbeing. In S. Smith, R. Pain, S. Marston and J. P. Jones III, eds., *The Sage handbook of social geographies.* London: SAGE Publications, pp. 308–29.

Kearns, R. A., Collins, D. and Neuwelt, P. M. 2003. The walking school bus: extending children's geographies? *Area,* 35(3), pp. 285–92.

Kearns R. A., Lewis, N., McCreanor, T. and Witten, K. 2009. 'The status quo is not an option': community impacts of school closure in South Taranaki, New Zealand. *Journal of Rural Studies,* 25, pp. 131–40.

Massey, D. 2004. Geographies of responsibility. *Geografiska Annaler,* 86B(1), pp. 5–18.

McKendrick, J. H. 2014. Geographies of children's well-being: in, of, and *for* place. In A. Ben-Arieh, F. Casas, I. Frones and J. Korbin, eds., *Handbook of child well-being: theories, methods and policies in global perspective.* Dordrecht: Springer, pp. 279–300.

Metcalfe, A. and Game, A. 2014. Ecological being. *Space and Culture,* 17(3), pp. 297–307.

Morrow, V. and Mayall, B. 2010. Measuring children's well-being: some problems and possibilities. In A. Morgan, E. Ziglio and M. Davies, eds., *Health assets in a global context: theory, methods, action.* New York: Springer, pp. 145–65.

Robbins, P. 2004. *Political Ecology: A Critical Introduction.* Malden, MA: Blackwell.

United Nations International Children's Fund (UNICEF). 2011. *Children and climate change: children's vulnerability to climate change and disaster impacts in East Asian and the Pacific.* Bangkok: UNICEF East Asian and Pacific Regional Office.

Wood, B. E. 2012. Crafted within liminal places: young people's everyday politics. *Political Geography,* 31(6), pp. 337–46.

Wood, B. E. 2013. Young people's emotional geographies of citizenship participation: spatial and relational insights. *Emotion, Space and Society,* 9(1), pp. 50–8.

Conclusions

The atmospheric attunements of children and young people

Stuart C. Aitken

Atmospheric attunements are not just the effects of a distant something elsewhere but the actual affects of modes of living being brought into being. A commonplace, labor-intensive process that stretches across imaginaries, social fields, sediments and airs, linking disparate and incommensurate registers and scales into some kind of everything. . . .The lived spaces and temporalities of home, work, school, blame, adventure, illness, rumination, pleasure, down time, release and phantasmatic or unthinkable situations are the rhythms of the present as a compositional event weighted with atmospheric fill.

Stewart, 2010: 6

For many years, I've been writing about the ways that the actions and activities of children and young people, and indeed their very presence, are affects that transform our modes of living. They are affects that are at once ordinary and transgressive, new and often wise, and perhaps also inspirational and liberatory. Children and young people are a large part of the quantitative and qualitative energy of our commonplaces and of our collective geographic imaginaries. Young people (under 20 years old) make up 35 percent of the world's population; they are disproportionately located in the Global South, contributing to the spreading bases of population pyramids, but in the Global North they also comprise the section of the population that raises concerns over future economic insecurities (Barr and Malik, 2016). Some policy makers raise the specter of the breakdown of children's health and increasing dependencies on environments and supporting social systems that are unable to cope, while others point to the increasing wage earning gap between young people and elders. These are all effects of the presence of young people, creating in their wake the moral panics and insecurities of the moment. There are also attendant hopes for the future, positionings and strategies focused on enduring and sustainable outcomes that reduce the burden on the world's youth. For all these reasons, and for better or worse, young people are now recognized as central actors in global politics, economics, and technologies. Since the UN Convention on the Rights of the Child (UNCRC, 1989, re-convened and reconstituted in 2009; UNICEF, n.d.), a growing consensus in the humanities, sciences, social sciences, public health, law and medicine suggests that young people must be understood as individuals and they must be recognized as members of a generational class with

248 Stuart C. Aitken

political, economic, social and cultural value and power. The last two decades witnessed a prolific rise in research on the ways healthy environments coincide, or not, with young people's geographies. This field has emerged by creating research synergies on the dynamics between economic, social, political, cultural, and natural environments as they apply to young people's livelihoods and wellbeing. The recognition of space as a critical context of human interaction has been significant in the humanities, sciences and social sciences for some time but a coherent multi-disciplinary focus on young people, their spatial relations, and healthy environments is only beginning to develop. This development is more than just about space and spatial relations, which are most often about what the world is; rather, our critical questions revolve around what the world, and the child as part of that world, does.

This book is one of the first to focus on youth geographies and health from a broad inter-disciplinary and transdisciplinary perspective. What it does well is bring together research on children, youth, and families as individual subjects as well as on the dynamics between them, their schools, communities, spaces, and other environmental factors at various scales to highlight the importance of creating a nexus of health and wellbeing. To do so, there is an implicit and often explicit understanding that space is a critical aspect of young people's lives and wellbeing. What is necessary and clear is the need to bring together the disciplines of geography, anthropology, history, environmental health, art, and child/family development and so forth in synergistic clusters that anticipate the future of research on young people and their environments. These synergies address the critical questions, many of which are well articulated and pondered on in the preceding pages: what does the world do for children and young people, and what are its health effects and affectations? What do children do through their agency and presence? How do affects show up in the commonplaces of healthy bodies and in their movements through and with other bodies, populations, environments, and institutions?

Kathleen Stewart (2010: 2) reminds us that the 'charged atmospheres of everyday life' are pernicious, malleable, and likely to snap in the most vulnerable places and at the most inopportune moments. She talks about the 'worlding of places . . . accreted out of what we might call opening events' (Stewart, 2010: 3). These may be toxic outbreaks, rude gestures, increased obesity, changes in political leadership, cancer clusters, or smiles and kind words that move through a group of young people that speed up or slow down, that exaggerate or settle. The sparser and more vulnerable life becomes, according to Stewart, the more these worldings accrue. Bodies are more susceptible, more engaged, when the world is vulnerable. The events that accrue, their intensities, become the air people breathe: racist violence, a dictatorship, sick children, a failing education system. Black boys join gangs to protect themselves from the police; fathers give up jobs to look after their ailing children; grandparents rent apartments in an unknown city so that their grandchildren can live with them and get a decent education; young girls heading north to the US border supply themselves with morning-after pills because they know they will be raped. Stewart calls the accrual of these events – violence, love, place, family, telling stories – the atmospherics with which we live because they are what we attend to, they are what we sense, and they are what we hope and imagine through. For young people they pose the possibility of resistance and transformation, of attachment and detachment, of abjection, of wellness, of anticipation, of staying and moving on, of constraints and possibilities, of becoming other than what they see in us adults. They are border crossings and revolutionary imaginations (Aitken *et al.*, 2011). What I want to do in this late chapter is think through the ideas in the preceding chapters in ways that help us understand what they mean for creating a healthy spatial nexus or assemblage for children and young people. I want to consider what matters in terms of this health, of these atmospherics, and what does not. What are their rhythms and temporalities? Where do they reside and how do they move? What is abusive and what is loving? What is a loss and what is a gain? For whom? When? Where? What, then, are the spatial resonance of the worldings and accruals? Can we intervene? Should we intervene? Do we know how to? What is the very best kind of atmospheric attunement for children and young people? This is a heady agenda around which, I warn you in advance, I will fail spectacularly. But, as I've written elsewhere, failing spectacularly is not necessarily a bad thing and it is important to note that experimentation and transdisciplinarity, as evidenced in many of the chapters in this book, is not for the timid or those who want tenure in those disciplinary enclaves called departments (Aitken, 2010: 220).

The research in this book is transdisciplinarity at its best. Unlike entrenched disciplinary perspectives, many of which are raised and connected to in the preceding pages (e.g. affordances, ecological theory ... more about these in a moment), or interdisciplinarity, which focuses on splicing and suturing perspectives from different disciplines, often in awkward and unwieldy ways, transdisciplinarity raises opportunities for taking us on academic adventures with the old to create something new and, maybe, revolutionary. The book is also disciplinary in the sense that several chapters build on the advantages of past work and move it forward with new empirical data. This is all good and appropriate. What binds it all together is an energy that incorporates aspects of young people's health and spaces in the humanities, and social and biological sciences. Among many other things, the work relates, and is relational, to children's environmental and embodied health, young people's mobility and access to resources, social media environments, youth citizenship and activism, restorative/therapeutic environments, and healthy communities. A disciplinary, multi-disciplinary, transdisciplinary, intertwined and relational focus on young people and their wellbeing is critically enhanced by the chapters *in toto*, with a coherent focus on space and spatial relations, and all this is important precisely because it anticipates the roles that children and youth play in shaping and changing the future, in creating and recreating themselves and their worlds.

Some chapters are very specific, systematically elaborating effects and parameters, or presenting new data to confirm old ideas. Some chapters think of space as a container of young people's activities, while others mince around specifics seeking to evoke rather than elaborate. For others, space is relational and an important actor with agency and affects. Others are trenchantly political, pushing a current moment where youth forms a primary vehicle for social movements, environmental activism, and information technologies. Insights from these varied approaches to the complexities of social and spatial systems help us to understand, at the very least, how and why certain places are healthy and therapeutic for young people while others are detrimental to their wellbeing. Some chapters, for example, focus on children and community environmental health in order to contribute to increased understanding of how environmental exposures vary over space, children's ages (including *in utero*), and how behavioral and cultural influences contribute to young people's exposures. The importance of considering the whole physical, developmental, economic, cultural, and social environment to study children's health and wellbeing is an area of emerging research in the sciences and social sciences and, to a larger and larger degree, in the humanities. Biologically, children and youth are at times among the most vulnerable to environmental exposures due to their developing bodies and behaviors (e.g. hand-to-mouth movements and increased physical activity that increases absorbed doses relative to adults) and they are often the most resilient, the quickest to bounce back. As several of the chapters in this book point out, interactions between factors not usually studied together can produce important insights. Neighborhood characteristics can, for example, effect exposures, such as access to parks and safe places to exercise, and collaborating with urban planning, architecture and geography is increasingly recognized as an important factor in promoting public health.

The social and emotional wellbeing of children, youth, and families with an emphasis on schools and communities is of critical importance to the atmospheric attunement of children and young people. Researchers in this book show up with expertise on early childhood mental health, violence prevention, and trauma-informed care that help focus projects on educational and support services for marginalized populations such as low-income Latino and African–American youth, children and parents in occupied Palestine, families in Ecuadorian shanty-towns, and early childhood education and mental health services for diverse populations. Throughout many of the chapters in this book, a focus on the social, emotional and ecological wellbeing of youth and families, youth leadership and education, and transformation in neighborhoods characterized by poverty, health disparities, and environmental exposures is aligned with work to better understand how social and spatial systems can foster health. Reported longitudinal research is rendered in both local and international contexts, and has provided a considerable depth of data on cross-cultural definitions of and relationships to childhood; the impacts of environmental toxins (e.g. lead) on physical and cognitive development; nutritional shifts as a result of changing food environments; advocacy related to the use of chemical additives in foods; the impact of poverty and political instability on youth migration; the role of folk models in shaping early childhood and K-12 educational curricula and settings; adaptations in familial and community construction due to migration and displacement; citizenship and new understandings of nationhood; and much more. Recognition of diversity in personal exposures has spurred an increased focus on exposure assessment at the level of the child, whether by examination of air young people breath at the home or school, contaminants in their soil and house dust, water, food, or by absorbed dose using biological markers (biomarkers) of exposure in blood, urine, or hair. Collaboration with researchers in other fields is increasingly necessary, bringing computer scientists, mobile sensing experts, geographers, and health professionals together with environmental health researchers. It is so important to measure a child's personal exposure to environmental contaminants. The research area can be global or local, but must focus on disparities in exposures. A track record of multi- and transdisciplinary collaboration is always desirable.

Children's mobilities are writ large in a number of chapters, and some are contextualized through the new mobilities literature. To the extent that a young person's spatial mobility is, as one chapter notes, about wellbeing affordances, it is about privileges, indulgences, and permissions, and it is also about what is possible in ways that give space agency and affect. Mobility and affordance, like competence, relates to concepts that arose from the beginnings of cognitive and environmental psychology in the 1970s and, in particular, the work of Roger Barker and James Gibson. The ecological perspectives of the time drew from Barker to seek understanding of the ways young people used environments in terms of which specific features had functional significance. Gibson elaborated the notion of affordances to describe the ways that environmental features resonated with the competencies (including health) of individuals. Thus, a tree is not a discrete and abstract object for a child, but it may have attributes (e.g. accessibility, climb-ability) that relate to the competencies and abilities of the child (e.g. size, hand-grasp, neuro-motor skills, balance), which in combination affords an opportunity for adventure and fun, or frustration. It is this complex set of relations that some of the researchers in this book are interested in, but not in the somewhat reductionist ways of ecological psychologists who followed Gibson and Barker. The work in this book pushes this literature beyond the developmental and ecological processes involved in how children learn to manipulate specific aspects of environments (affordances) to pay more attention on the ways that those spaces are constructed and opened physically, imaginatively, and emotionally, and through belief systems, values, and ideologies.

Geographies of health are a dynamic and increasingly important aspect of the sciences, social sciences, and humanities to the degree that they focus on the multi-faceted relationships between health and places, including research on demographic patterns and ecologies of disease, global processes of disease transmission, environmental racism, healthy neighborhoods, therapeutic landscapes, food sustainability, and more. In some chapters there is support for an ecological approach to understanding of lifespan development with the focus on infants, children, middle childhood, adolescents, and young adults but these are always tempered with strong research methods, knowledge, and expertise.

The fusion of epidemiology, the ecological approach and the evolutionary science of adaptation enables deeper understandings of both the symbolic and attuned experiences of young people. The fusion provides writers in this book critical insights into local processes and practices. A critical area of study on youth includes the impact of mobilities and technologies on the modification of life's atmospherics. What is the impact of mobility on children who fall ill and need specific medical treatments? What are the physical, social, emotional, and health dimensions of families apart, or of children moved to specialized treatment centers? With evidence for increased global movement in search of appropriate medical treatments, what happens when children are torn from the loving and caring environments of family and friends? What happens when they lose parents or siblings to these movements? What kind of burden is placed on local and international health care systems? A greater understanding of how children and youth are affected by and respond to these spatial relations and how they modify existing practices and traditions to create new atmospherics is imperative for open political, economic, health, and environmental processes. Mobilities show up also in rural, suburban, and urban places, posing questions that relate to young people's physical activities, to their living in food deserts, or their access to medical facilities. The degree of constraints on young people's mobilities within cities and across borders is another critical area of health geographies outlined in the preceding pages. In some contexts, independent mobility is encouraged as a boon to healthy living; in others it is a liability or dangerous to children's health. Questions revolve around relations with nature, contexts of citizenship, and contexts of life. How does technology affect young people's mobilities? Where do mobile communications move young people? Do they move them closer to family members and friends? Does the family nexus become more cybernetic? Do mobile communications foment more rules or more privileges? Do they increase agency? With focus on a moving blue dot in a phone, is there less concern about the world and immersion in it? Are young people returning to a pre-Copernican world-view? These and other critical questions are addressed in this book pursuant of the finding that the health of children is unequivocally connected to the health of the planet, and the composition of appropriate atmospherics, perhaps even happy atmospherics.

As I write this chapter, the news is filled with a series of environmental disasters involving children. The Zika virus—impeding brain development in utero—explodes through and beyond Brazil; in the US, it seems that years of lead-poisoning of children in Flint, Michigan was the result of government cover up at the local and federal level. In both these cases, the impact is felt most heavily by poor children and the burden of responsibility, with the exception of the rolling heads of a few officials, is borne by their poor parents. Increasingly, it seems that there is no 'right to care' about the health of our children (to borrow Stephen Healy's (2008) concept). Children, carers, and health providers alike are increasingly stuck in a market-driven neoliberal system in which rational self-interest guarantees lack of provision and scarcity as the ontological grounds for the healthcare economy. This market-driven system assumes that patients who are covered by insurance consume as much care as possible, while providers' self-interest pushes the diagnosis and treatment of illness to maximize income. Children in Brazil and Flint become victims of a care system that does not care. The only way forward, it seems, is a problematic and debilitating push for measures that remove culpability, reduce costs and limit the use of resources. As a consequence, cost savings push care out of institutional environments and towards a redistribution of labor that places the bulk of after-care and health maintenance on the patient and their system of informal caregivers, and on the child and their extended family.

Although there is outrage over what happened in Flint, and the world seems to mobilize in support of eradicating the mosquitoes that carry the Zika virus, these are all appearances and media representations that do little to change an underlying system that increasingly burdens the responsibility of care on those who are least able to bear it. Henri Lefebvre (2004) argues the importance of approaching the world in terms of what it evokes, what it brings into being, and what is does, rather than for what it is, what it appears to be. With the chimera of appearances in mind, Stewart (2010: 6) argues further that the world demands to be heard in a meaningful way; and, as such, it demands engagement and tuning in. Not to tune in, or to abstract our engagement with just words, theories, models, and mathematics, loses our opportunity to connect with Stewart's atmospherics. There is nothing wrong with words, theories, models, and mathematics and, like everything else, they have intended and unintended consequences. Clearly, nothing is fully intended or unintended; there are always surprises and openings, unsought heroes, abandonments and reconciliations, foreclosures and emancipations. The push since the 1989 UNCRC for young people's participation and agency is problematic because there is never unadulterated agency nor is there ever pure passivity, but there is always the presence of young people. Noting the presence of children, which the preceding chapters do so well, is our first key action because to dismiss or deny presence is in-and-of-itself not only unhealthy, it is a form of oppression. The next key is to remain open and attuned to the possibilities, to not foreclose upon the political and the hopeful (Massey, 2005), and to know that the possibilities opening up are not always good ones . . . but they may be (Stewart, 2010: 6). We cannot promise children and young people a better, healthier future with any certainty, but we can open it up to good and healthy possibilities. To hide behind what worked in the past (the model that predicts well or the theory that unravels spatial injustice precisely) is to create frames that foreclose upon the political. Spatial frames and representations are not good places within which to be stuck. To be stuck is to lose potential, and so it is important to appreciate that

> . . . [a]n atmospheric fill resonates the edge between the material and the potential. As a proliferative condition it not only allows, but spawns, the production of different worlds, experiences, conditions, dreams, imaginaries and moments of hyperactivity, down time, interruption, flow, friction, eruption, and life styles.
>
> Stewart, 2010: 8

In our push to create a nexus of health, it is important to note that every (young) person is a nexus of compositional moments that are relational and shared. We need to rethink our analytic and critical processes to include attunement to openings and surprises, we need to jump in intuitively, as suggested years ago by Henri Bergson, to atmospheres as living, pulsing, moving compositions.

References

Aitken, S. C. 2010. Bold disciplinarianism, experimentation and failing spectacularly. *Children's Geographies*, 8(2), pp. 219–20.

Aitken, S. C., Swanson, K., Bosco, F. and Herman, T. 2011. *Young people, border spaces and revolutionary imaginations*. New York: Routledge.

Barr, C. and Malik, S. 2016. Revealed: the 30-year economic betrayal dragging down Generation Y's income. *The Guardian*, Monday 7 March.

Healy, S. 2008. Caring for ethics and the politics of health care reform in the United States. *Gender, Place and Culture*, 15(3), pp. 267–84.

Lefebvre, H. 2004. *Rhythmanalysis*. London: Continuum.

Massey, D. 2005. *For space*. New York: Routledge.

Stewart, K. 2010. Atmospheric attunement. *Rubric: Writing from UNSW*, 1, pp. 2–14.

United Nations International Children's Emergency Fund (UNICEF). n.d. Convention on the Rights of the Child. Available at https://www.unicef.org/crc/ (accessed 26 February 2017).

Index

Page numbers in italics refer to illustrations and tables.

Milton Keynes UK
Ingram Content Group UK Ltd.
UKHW040109071024
449327UK00019B/933